SPACE SCIENCE, EXPLORATION AND POLICIES

THE INTERNATIONAL SPACE STATION

MANAGEMENT AND UTILIZATION ISSUES FOR NASA

SPACE SCIENCE, EXPLORATION AND POLICIES

Additional books in this series can be found on Nova's website under the Series tab.

Additional e-books in this series can be found on Nova's website under the e-book tab.

SPACE SCIENCE, EXPLORATION AND POLICIES

THE INTERNATIONAL SPACE STATION

MANAGEMENT AND UTILIZATION ISSUES FOR NASA

EVELYN CLEMENS
EDITOR

nova publishers
New York

NOTICE TO THE READER

The Publisher has taken reasonable care in the preparation of this book, but makes no expressed or implied warranty of any kind and assumes no responsibility for any errors or omissions. No liability is assumed for incidental or consequential damages in connection with or arising out of information contained in this book. The Publisher shall not be liable for any special, consequential, or exemplary damages resulting, in whole or in part, from the readers' use of, or reliance upon, this material. Any parts of this book based on government reports are so indicated and copyright is claimed for those parts to the extent applicable to compilations of such works.

Independent verification should be sought for any data, advice or recommendations contained in this book. In addition, no responsibility is assumed by the publisher for any injury and/or damage to persons or property arising from any methods, products, instructions, ideas or otherwise contained in this publication.

This publication is designed to provide accurate and authoritative information with regard to the subject matter covered herein. It is sold with the clear understanding that the Publisher is not engaged in rendering legal or any other professional services. If legal or any other expert assistance is required, the services of a competent person should be sought. FROM A DECLARATION OF PARTICIPANTS JOINTLY ADOPTED BY A COMMITTEE OF THE AMERICAN BAR ASSOCIATION AND A COMMITTEE OF PUBLISHERS.

Additional color graphics may be available in the e-book version of this book.

Library of Congress Cataloging-in-Publication Data

ISBN: 978-1-63484-071-2

Published by Nova Science Publishers, Inc. † New York

CONTENTS

PREFACE

The U.S. has spent almost $43 billion to develop, assemble, and operate the International Space Station (ISS) over the past two decades. The NASA Authorization Act of 2010 required NASA to enter into a cooperative agreement with a not-for-profit entity to manage the ISS National Laboratory and in 2011 did so with the Center for the Advancement of Science in Space (CASIS). CASIS is charged with maximizing use of the ISS for scientific research by executing several required activities. Recently, questions have arisen about the progress being made to implement the required activities and the impact it has had on ISS's return on the investment. This book assesses the extent to which CASIS has implemented the required management activities; and NASA and CASIS measure and assess CASIS's performance. Furthermore, this book assesses the extent to which NASA has ensured essential spare parts are available and ISS structures and hardware are sound for continued ISS utilization through 2020.

In: The International Space Station
Editor: Evelyn Clemens
ISBN: 978-1-63484-071-2

Chapter 1

International Space Station: Measurable Performance Targets and Documentation Needed to Better Assess Management of National Laboratory*

United States Government Accountability Office

Why GAO Did This Study

The U.S. has spent almost $43 billion to develop, assemble, and operate the ISS over the past two decades. The NASA Authorization Act of 2010 required NASA to enter into a cooperative agreement with a not-forprofit entity to manage the ISS National Laboratory and in 2011 did so with CASIS. CASIS is charged with maximizing use of the ISS for scientific research by executing several required activities. Recently, questions have arisen about the progress being made to implement the required activities and the impact it has had on ISS's return on the investment.

GAO was asked to report on the progress of CASIS's management of the ISS National Laboratory.

* This is an edited, reformatted and augmented version of a United States Government Accountability Office publication, No. GAO-15-397, dated April 2015.

GAO assessed the extent to which (1) CASIS has implemented the required management activities, and (2) NASA and CASIS measure and assess CASIS's performance. To perform this work, GAO reviewed the cooperative agreement between NASA and CASIS, CASIS's annual program plans, and other documentation and interviewed ISS, CASIS, and NASA officials.

WHAT GAO RECOMMENDS

GAO recommends NASA fully staff the ISS National Laboratory Advisory Committee; NASA and CASIS work together to develop measurable targets for CASIS's metrics; and NASA begin documenting its annual review of CASIS's performance. NASA partially concurred and CASIS did not concur with the first recommendation, but concurred with the other two. GAO continues to believe the first recommendation is valid, as discussed further in the report.

WHAT GAO FOUND

The Center for the Advancement of Science in Space (CASIS), manager of the International Space Station (ISS) National Laboratory, has taken steps to fulfill its management responsibilities contained in its cooperative agreement with the National Aeronautics and Space Administration (NASA), and has initiated the activities required by the NASA Authorization Act of 2010. GAO found that CASIS implemented procedures for prioritizing research; evaluated 206 proposals and awarded approximately $20 million in grants to 77 research projects through January 2015; and cultivated relationships with academic institutions, research-specific organizations, and other entities. CASIS, however, has not been able to fulfill its responsibility in the cooperative agreement to interact with the ISS National Laboratory Advisory Committee, which NASA was statutorily required to establish under the Federal Advisory Committee Act, because NASA has yet to staff the committee as required by the NASA Authorization Act of 2008. As a result, CASIS is not able to fulfill its responsibility in the cooperative agreement that requires it to coordinate with this committee and review any report or recommendations it originates.

CASIS has established fiscal year 2015 metrics that meet most of GAO's key attributes for successful performance measures; however, NASA and CASIS did not establish measurable targets for these performance metrics, and NASA's annual assessment of CASIS was not documented.

GAO's work on best practices for measuring program performance has found that performance metrics should have quantifiable targets to help assess whether goals and objectives were achieved by easily comparing projected performance and actual results. CASIS officials told GAO in July 2014 that setting measurable targets would be arbitrary because CASIS processes and metrics are still evolving. In January 2015, however, the Chairman of the CASIS Board of Directors told GAO that setting measurable targets is a priority for the board. CASIS, however, has yet to establish a date by which measurable targets will be developed. Using the established metrics, NASA is required by the cooperative agreement to perform an annual program review of CASIS's performance. This review is informal and not documented as ISS program officials provide the results to CASIS orally. This approach is inconsistent with federal internal control standards, which call for information to be recorded and communicated to those who need it to manage programs, including monitoring performance and supporting future decision making. Although NASA officials reported that they were generally satisfied with CASIS's performance, CASIS officials said a formal summary of the results would make the information more actionable.

ABBREVIATIONS

CASIS	Center for the Advancement of Science in Space
CATO	Cooperative Agreement Technical Officer
INLAC	ISS National Laboratory Advisory Committee
ISS	International Space Station
NASA	National Aeronautics and Space Administration
RFP	request for proposal
SpaceX	Space Exploration Technologies Corporation
STAP	Science and Technology Advisory Board
STEM	science, technology, engineering, and mathematics

April 27, 2015

The Honorable Lamar Smith
Chairman
Committee on Science, Space, and Technology
House of Representatives

Dear Mr. Chairman:

The International Space Station (ISS) has been used as a manned research outpost continuously for over 14 years. Much of the National Aeronautics and Space Administration (NASA) ISS research has focused on reducing the known risks related to long-duration human spaceflight as the agency plans for missions beyond low-Earth orbit. The United States has spent over $43 billion to develop, assemble, and operate the ISS over the past two decades and NASA plans to spend about $17 billion over the next 5 years to enable further scientific research the agency views as critical to future human space activities. In January 2014, the Administration proposed extending the life of the ISS by a minimum of 4 years to at least 2024. To take further advantage of the investment in the ISS, Congress enacted several laws to increase utilization of the ISS by commercial and academic researchers, first with the designation of the ISS as a National Laboratory, and then later by directing NASA to enter into an agreement with a not-for-profit entity to manage the ISS National Laboratory.[1] In 2011, NASA selected the Center for the Advancement of Science in Space (CASIS), a non-profit entity, to manage non-NASA commercial and academic research aboard the ISS National Laboratory.

In March 2014, you stated that the continued operations of this unique research facility was a priority for the committee, but also expressed concern about the progress being made by CASIS to manage the facility. You requested that we review and report on CASIS's progress in managing the ISS National Laboratory, including a review of the execution of the activities required by the NASA Authorization Act of 2010. We assessed the extent to which (1) CASIS has initiated and implemented the required management activities for non-NASA research aboard the ISS National Laboratory, and (2) NASA and CASIS measure and assess CASIS's performance.

To assess the extent to which CASIS has implemented required management activities for non-NASA research aboard the ISS National Laboratory in the NASA Authorization Act of 2010, we reviewed the cooperative agreement between NASA and CASIS and the fiscal year 2014 and 2015 Annual Program Plans. We also examined documentation on CASIS's portfolio management, research prioritization and proposal review and evaluation processes, business development efforts and outreach efforts, and education initiatives.

To obtain additional information on CASIS's implementation of some of the required activities, we conducted structured interviews with 14 of 20 randomly sampled researchers who had submitted proposals to CASIS through July 2014. While our sample was, in part, representative of the population of 172 researchers, the results from the relatively small sample size are not generalizable to the population. To determine the extent NASA and CASIS measure and assess CASIS's performance, we evaluated CASIS's fiscal year 2014 and 2015 metrics to determine whether they adhered to GAO's key attributes of successful performance measures. We also reviewed CASIS documentation related to measuring and reporting the results of these metrics.

We reviewed the cooperative agreement between NASA and CASIS to determine relevant NASA responsibilities. In addition, we interviewed NASA and CASIS officials to understand the processes involved with developing CASIS's metrics and assessing its performance. See appendix I for a detailed description of our scope and methodology.

We conducted our review from April 2014 to April 2015 in accordance with generally accepted government auditing standards. These standards require that we plan and perform the audit to obtain sufficient, appropriate evidence to provide a reasonable basis for our findings and conclusions based on our audit objectives. We believe that the evidence obtained provides a reasonable basis for our findings based on our audit objectives.

BACKGROUND

The ISS supports research projects with state of the art facilities for Earth and space science, biology, human physiology, physical science, and materials research, and provides a platform to demonstrate new space-related technologies.

Source: NASA. | GAO-15-397.

Figure 1. The International Space Station.

The facilities include modular multipurpose payload racks and external platforms to store and support experiments, refrigerators and freezers for biological and life science samples, research lockers or incubators, and a combustion chamber to observe combustion patterns in microgravity, among other research equipment. The ISS currently has three crew members in the U.S. operating segment who, according to NASA officials, devote a total of approximately 35 hours per week to conduct research. The remaining crew time is used for operations and maintenance of the ISS, training, exercise, and sleep. NASA plans to increase the number of astronauts in the U.S. operating segment of the ISS from three to four once a U.S. capability to transport crew to and from the ISS is available.[2] Cargo transportation to the ISS is done through a commercial resupply services contract that was signed with Orbital Sciences Corporation (Orbital) and Space Exploration Technologies Corporation (SpaceX) in 2008. SpaceX currently has a capsule that can also return significant amounts of cargo to Earth and is the only vehicle currently servicing the ISS that has this capability. Orbital and SpaceX are scheduled to provide 8 and 15 resupply flights, respectively, through December 2017. As of January 2015, SpaceX has launched five successful resupply missions and Orbital has launched two successful resupply missions. Orbital resupply flights to the ISS were deferred pending a review of a mishap that occurred during a resupply launch in October 2014, which resulted in the loss of that mission. According to NASA officials, a "return to flight" plan was submitted by Orbital and accepted by the ISS program in January 2015.

Legislation on Management of the ISS National Laboratory

Since 2005, Congress has directed several changes regarding the management and utilization of the ISS. The NASA Authorization Act of 2005 designated the U.S. segment of the ISS as a National Laboratory.[3] The 2005 act directed the NASA Administrator to seek to increase ISS utilization by other federal entities and the private sector through partnerships, cost-sharing agreements, and other arrangements that would supplement NASA funding of the ISS. It also allowed the Administrator to enter into a contract with a nongovernment entity to operate the ISS National Laboratory.

The NASA Authorization Act of 2008 further directed NASA to establish the ISS National Laboratory Advisory Committee, which was to be composed of individuals representing organizations that had formal agreements with NASA to utilize the U.S. portion of the ISS. [4] The 2008 act stated that the committee shall monitor, assess, and make recommendations regarding effective utilization of the ISS as a national laboratory and platform for research, and submit a report containing these assessments and recommendations at least annually to the NASA Administrator.

The NASA Authorization Act of 2010 required the NASA Administrator to provide initial financial assistance and enter into a cooperative agreement with a not-for-profit organization to manage the activities of the ISS National Laboratory for non-NASA utilization of the ISS research capabilities and available facilities.[5]

The Administrator was required to designate a NASA liaison, with whom the selected not-for-profit entity would cooperate and consult with in carrying out its responsibilities under the agreement. An individual in the Space Life and Physical Sciences Research and Applications Division of the Human Exploration and Operations Mission Directorate is currently serving as the NASA liaison. The 2010 act outlined seven management and research and development activities that NASA was required to provide funding for the not-for-profit entity to initiate.[6] Those activities stated briefly, are to:

- Plan and coordinate ISS National Laboratory research activities;
- Develop and implement guidelines, selection criteria, and flight support requirements for non-NASA utilization of the ISS research capabilities and available facilities;
- Interact with the ISS National Laboratory Advisory Committee and review recommendations provided by that committee;
- Coordinate transportation requirements in support of the ISS research and development objectives;
- Cooperate with NASA, other departments and agencies of the U.S. government, and commercial entities to sustain ground support facilities for the ISS;
- Develop and implement scientific outreach and education activities designed to ensure effective utilization of ISS research capabilities; and
- Address other matters relating to the utilization of the ISS National Laboratory for research and development as the Administrator may consider appropriate.

The 2010 act also requires the ISS National Laboratory-managed experiments to be guaranteed access to and use of at least 50 percent of the U.S. research capacity allocation including power, facilities to keep experiments cold, and requisite crew time onboard the ISS through September 30, 2020.

The Administrator can allocate additional capacity to the ISS national laboratory if this capacity is in excess of NASA research requirements. If any NASA research plan requires more than the at least 50 percent of the U.S. research capacity allocation of ISS resources, the plan should be submitted for consideration of proposed research to be conducted within the ISS National Laboratory capacity of ISS resources. The person designated as the NASA liaison to the not-forprofit entity has the authority to provide those resources beyond its 50 percent allocation on an exception basis if a proposed experiment is considered essential for purposes of preparing for exploration beyond low-Earth orbit, based on a joint agreement between the NASA liaison and the not-for-profit entity.

CASIS and NASA Responsibilities Outlined in Cooperative Agreement

In August 2011, after a competitive process, NASA signed a cooperative agreement with CASIS, a not-for-profit entity, to manage the activities of the ISS National Laboratory through September 30, 2020.[7] Cooperative agreements differ from contracts. Generally, cooperative agreements are used when the principal purpose of a transaction is to stimulate or support research and development for a public purpose, and substantial involvement is expected between the executive agency and the award recipient when carrying out the activity identified in the agreement. In contrast, contracts are used when the principal purpose is acquisition of property or services for the direct benefit or use of the federal government.[8]

CASIS is bound by the responsibilities outlined in the cooperative agreement, which tasks CASIS with maximizing the value of the ISS National Laboratory by stimulating interest and use of the ISS for scientific research by directly soliciting potential users and fostering a market to attract others. CASIS is also charged with maximizing the use of the ISS for advancing science, technology, engineering, and mathematics (STEM) education. Pursuant to the cooperative agreement, NASA will provide CASIS $15 million annually through 2020, of which it will seek to award at least $3 million for research grants. CASIS officials said that the remainder of NASA funding is used for infrastructure and direct costs such as labor and travel-related expenses. According to the cooperative agreement, CASIS will solicit non-NASA funding for research by targeting various sources—such as government grants, foundation funding, charitable contributions, private equity, venture financing, and private investors—and facilitate matching of projects that meet the research objectives with those qualified funding sources. Additionally, the cooperative agreement requires the development of an annual program plan, which includes a detailed plan of CASIS's proposed activities for the following year, which CASIS must meet using its "best efforts," and annual and quarterly performance metrics.

The cooperative agreement outlines responsibilities for NASA such as providing resources and accommodations to CASIS to meet ISS National Laboratory requirements and performing the payload operations integration to ensure safe and effective flight readiness and vehicle integration. The Cooperative Agreement Technical Officer, a NASA employee within the ISS Program Office at Johnson Space Center, is charged with oversight of the cooperative agreement. The Cooperative Agreement Technical Officer is to coordinate the approval of the Annual Program Plan and track performance to the plan using the metrics reflected in CASIS's quarterly reports.

CASIS IMPLEMENTED MOST OF THE REQUIRED ACTIVITIES, BUT IS NOT ABLE TO INTERACT WITH ADVISORY COMMITTEE

CASIS has taken steps to carry out its responsibilities to manage and promote research activities on the ISS National Laboratory as outlined in its cooperative agreement. For example, CASIS identified key research areas and released seven requests for proposals to solicit interest for research projects. Our survey of a sample of researchers who had submitted proposals to CASIS revealed generally positive comments about CASIS's management effort. For example, many respondents indicated that CASIS's processes were clear and that it evaluated their proposals fairly. CASIS, however, has not been able to coordinate with the ISS National Laboratory Advisory Committee (INLAC), as required, because NASA has yet to staff the committee.

CASIS Took Steps to Implement Management Activities

CASIS has taken steps to fulfill its responsibilities contained in its cooperative agreement with NASA, and has initiated the activities required by the NASA Authorization Act of 2010.[9] Table 1 summarizes the activities contained in the 2010 act as well as the corresponding responsibilities for CASIS and NASA outlined in the cooperative agreement.[10]

Table 1. Required Management Activities in the NASA Authorization Act of 2010 and Corresponding Responsibilities in the Cooperative Agreement

Required activities outlined in the NASA Authorization Act of 2010	The Center for the Advancement of Science in Space (CASIS) and NASA responsibilities outlined in the Cooperative Agreement
Plan and coordinate research activities	CASIS is to determine the research objectives that provide the most value to the nation across research disciplines and range of basic to applied research.

Develop and implement guidelines, selection criteria, and flight support requirements	CASIS is to prioritize the entire National Laboratory research portfolio using a fair, transparent, and impartial selection process that maximizes value of the International Space Station (ISS) investment using economic valuation techniques. NASA is to perform the activities to address flight support requirements.
Coordinate transportation requirements in support of the ISS National Laboratory	CASIS is to plan and coordinate the National Laboratory research activities for both ground and on-orbit execution and participate in the NASA research process. NASA will provide resources and accommodations to meet ISS National Laboratory requirements. NASA will perform the payload physical, analytical, andoperations integration during pre-flight, post-flight, transportation and on-orbit phases of the payload to ensure safe and effective flight readiness and vehicle integration.
Cooperate with NASA, other departments and agencies of the U.S. Government and commercial entities to ensure the enhancement and sustained operations of payload ground facilities	CASIS will develop the appropriate partnerships with other U.S. government agencies, academic institutions, and private firms and leverage implementation partners. CASIS will take into account appropriate research and technology development objectives, capabilities of the parties to perform the work required to select and execute research on the ISS, and the ability to provide financial sponsorship.
Develop and implement scientific outreach and education activities designed to ensure effective utilization of ISS research capabilities.	The CASIS Board and staff will engage in a deliberate, targeted advocacy campaign to stimulate interest in using the ISS National Laboratory for the conduct of research, technology demonstrations, and as a platform for science, technology, engineering, and mathematics education.
Interaction with and integration of the International Space Station National Laboratory Advisory Committee (INLAC).	CASIS will coordinate with the ISS National Laboratory Advisory Committee as established under section 602 of the NASA Authorization Act of 2008 (42 U.S.C. 17752) and review recommendations provided by the INLAC.

Source: GAO analysis of the NASA Authorization Act of 2010 and the Cooperative Agreement between NASA and CASIS. | GAO-15-397

Plan and Coordinate Research Activities

To determine its research and technology development objectives in accordance with the cooperative agreement, CASIS identified and prioritized the most promising research areas—which CASIS refers to as pathways—with guidance from the Science and Technology Advisory Panel, a CASIS committee comprised of both academic and commercial experts. These pathways are identified by compiling a list of research categories and determining financial feasibility. According to CASIS and NASA officials, research pathways are generated from various sources such as the Decadal surveys—studies conducted once every decade by the National Research Council that prioritize the research focus for the next 10 years in various scientific disciplines—and past NASA studies. To date, CASIS identified protein crystal growth, stem cell research, materials science, enabling technology to support science in space, Earth imaging, and remote sensing as key research pathways and developed a request for proposals (RFP) for each of these research pathways. CASIS released seven RFPs since it was established with the first occurring in June 2012, about 10 months after it was established. CASIS also accepts unsolicited proposals from researchers and other sources such as partnership accelerators and competitions.[11] As of January 2015, CASIS had received 206 proposals from all sources and awarded approximately $20 million in research grants to 77 projects, and paid almost $13 million to the awarded grants. Table 2 shows information related to the types of proposals CASIS has received and the number of grants awarded.

CASIS-sponsored research investigations awarded through its first RFP in 2012—involving protein crystal growth and microgravity—flew to the ISS National Laboratory in April 2014 and were returned to Earth in October 2014. These research investigations are currently in post flight analysis. As of December 2014, there were 11 CASIS-sponsored research investigations being conducted aboard the ISS National Laboratory.

According to NASA and CASIS officials, as CASIS increases the number of experiments for the ISS National Laboratory, the demand for crew time and certain research facilities aboard the ISS is expected to increase and they project the ISS National Laboratory will be challenged with meeting that demand. NASA officials explained that while the demand for crew time is currently manageable, it remains allocated at or near 100 percent, as the three crew members on the U.S. segment of the ISS utilize most of the 35 hours scheduled per week to conduct research. Crew time is expected to double on the ISS National Laboratory once the crew increases from three to four astronauts in fiscal year 2018 because, according to NASA officials, the additional crew member would be able to devote most of his or her time to research. NASA officials stated they are also working with CASIS to build automation into research experiments to reduce the monitoring time by crew members. Both CASIS and NASA expect increased demand for facility resources such as the Animal Enclosure Module used for rodent research and the remote sensing cameras used for Earth observation. Sharing of the ISS National Laboratory facilities requires considerable communication and agreement. NASA and CASIS officials said both organizations have ongoing discussions about how to share resources, coordinate research and ensure all users are represented when meeting the demand for crew time and ISS National Laboratory facilities and hardware. NASA officials explained that they reprioritize as necessary to ensure resources are not overstressed.

Table 2. Research Proposals Approved by the Center for the Advancement of Science in Space (CASIS) Since 2012

Solicited request for proposals (RFP)	Date RFP issued	Number ofgrants awarded
1. Advancing protein crystallization using microgravity	June 26, 2012	6
2. Materials testing in the extreme environment of space	September 4, 2012	2
3. Impact of microgravity on fundamental stem cell properties: a call for spaceflight and ground-based experiment	May 22, 2013	7
4. Remote sensing from the International Space Station	January 13, 2014	5
5. Enabling technology to support science in space for life on Earth	February 26, 2014	3
6. Materials science in space	April 28, 2014	3
7. Earth observation to benefit energy technologya	October 14, 2014	TBD
Other proposals		
Unsolicited proposals[b]		40
Accelerators and competitions[c]		11
Totals		77

Source: GAO presentation of CASIS data. | GAO-15-397

[a]Responses to this request for proposal were still being evaluated as of January 2015.

[b]Unsolicited proposals cover various topics and are not specific to a request for proposal.

[c]Accelerators and competitions—grants awarded through CASIS partnership with third parties that help to accelerate business development—are not specific to a request for proposal.

Develop and Implement Guidelines, Selection Criteria, and Flight Support Requirements

To initiate the development of guidelines and selection criteria, CASIS implemented procedures for prioritizing research, guidelines for proposal development, and evaluation and selection of research proposals, in accordance with the cooperative agreement.

- **Procedures for prioritizing research:** CASIS has implemented a multi-layer review process to identify and develop the overall research portfolio and prioritize future research pathways. See figure 2 for the process CASIS follows to prioritize research pathways.
- **Guidelines on proposal development and flight support requirements:** CASIS established guidelines that are incorporated in the applicable RFP for researchers to follow as they develop their proposals. The RFPs include specific criteria for proposal that CASIS uses as a basis for initial acceptance or denial of proposal submissions. For example, one RFP issued in 2014 contained minimum eligibility criteria such as the research being flight ready within 12 months of award and the research having secured funding, and included provisions that excluded the use of new sensors or instruments for remote sensing, and required that selected proposals be completed by 6 months post-flight. Each RFP also had unique criteria which can be dependent upon the research pathway and the facilities available on the ISS National Laboratory in the proposed time line. CASIS has separate guidance for unsolicited proposal submissions.

CASIS has documented the specific activities for meeting flight requirements, which includes the role of implementation partners and NASA in meeting these requirements. Implementation partners are subcontractors to CASIS and specialize in aerospace technologies and services. They have an integral role in providing hardware, flight integration services, and ground services to support CASISsponsored research. NASA performs the activities necessary to incorporate the research on a flight vehicle, such as providing the resources and accommodations to meet ISS National Laboratory requirements, and managing launch operations through payload return to Earth.

- **Evaluation and selection of research proposals:** CASIS implemented a policy that documents the submission and general review process for solicited and unsolicited proposals as well as proposals that it evaluates as part of agreements with outside organizations such as partnerships or subcontracts. The process begins with a two-step initial submission review for solicited proposals and a preliminary review for unsolicited proposals, then a five-step evaluation process. Figure 3 details the CASIS proposal evaluation process.

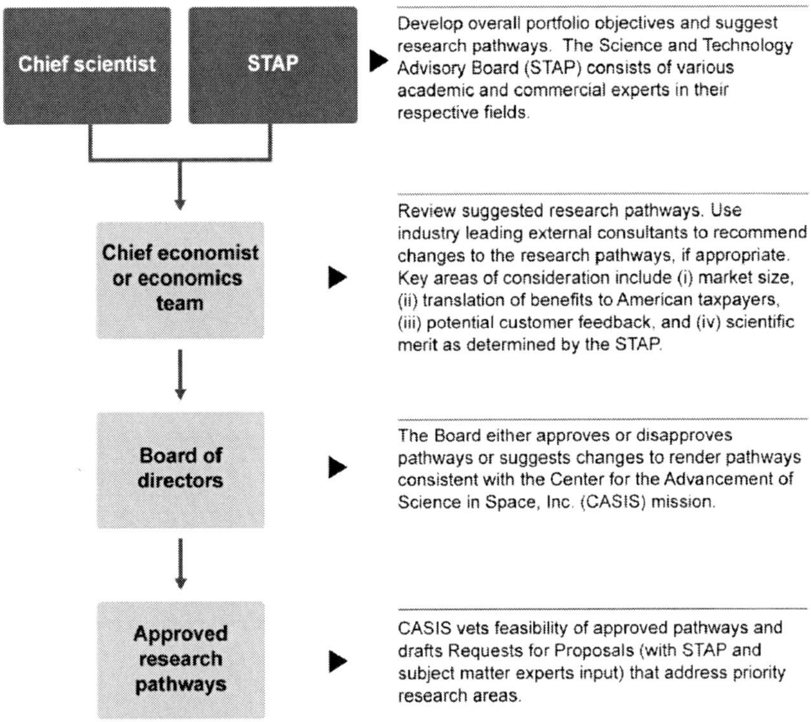

Source: GAO presentation of CASIS data. | GAO-15-397.

Figure 2. The Center for the Advancement of Science in Space (CASIS) Research Pathway and Development Process.

Proposal type	Solicited proposal	Unsolicited proposal
Initial proposal submission evaluation		
Preliminary submission	**Step-1 proposal:** The Center for the Advancement of Science in Space, Inc. (CASIS) Science team reviews an abbreviated proposal with project summary information that includes relevance to the CASIS RFP and goals.	The CASIS Science, Operations, and Fundraising (if appropriate) teams perform a preliminary review of a white paper or full proposal.
Full proposal submission	**Step-2 proposal:** Selected Step-1 proposals are invited to submit Step-2 proposals that contain more detail about the proposed project than is required in the first step. These Step-2 proposals move to the five-step evaluation process. If appropriate, CASIS Education and/or Development teams will review these proposals as well.	Based on feedback from the preliminary review, a proposer may expand and/or improve the proposal prior to the full CASIS review. Qualified unsolicited proposals then move forward to the five-step evaluation process. If appropriate, CASIS Education and/or Development teams will review these proposals as well.
Five step evaluation process for solicited and unsolicited proposals		
Operations review	The CASIS Operations team conducts a technical feasibility review of the proposal to ensure payload readiness/feasibility for flight.	
Scientific review	An external panel of subject matters experts assembled by CASIS evaluates the proposal for scientific merit and potential impact.	• If a proposal is requesting less than $100,000 in CASIS funding, the internal CASIS Science team completes a scientific review and engages external subject matter expertise only as needed. • If a proposal is requesting $100,000 or more, external subject matter experts complete the scientific review.
Economic review	CASIS economic staff, engaging additional subject matter expertise only as needed, performs an economic review of the proposal.	
Compliance review	The CASIS Compliance team assesses the proposal for regulatory and legal risks.	
Final determination	The Executive Director, Chief Scientist, and Chief Economist perform the final prioritization and award determination based on recommendations from the Operations/Science/Economic review teams and CASIS management-level staff as necessary.	

Source: GAO presentation of CASIS data. | GAO-15-397.

Figure 3. The Center for the Advancement of Science in Space (CASIS) Proposal Evaluation Process.

We surveyed a random sample of 14 researchers who submitted proposals to CASIS from 2012 through 2014 to obtain their perspectives on CASIS's performance in this and other areas. Although the results of this survey are non-generalizable because of our small sample size, overall the respondents were generally positive about their interaction with CASIS. For example, 11 of the 14 respondents indicated that CASIS's evaluation criteria were clearly articulated and 12 of 14 respondents believed their proposals were evaluated fairly. Of the 14 respondents, 13 said they were likely to submit future proposals to CASIS. In addition, all 14 respondents indicated that they were notified in a timely manner of the disposition of their proposal. CASIS declined proposals for 8 respondents. Of these 8, 7 said that they were provided feedback concerning why their proposal was declined. Several respondents, however, said that they were provided only a short bulleted response that fell short of addressing the scientific merit of the proposal. One respondent said they received a letter summarizing reviewers' comments that had several good points and was fair, but it was not detailed as it contained less information than what other grantors provide. According to CASIS guidelines, researchers whose projects are not selected for award are provided feedback and, are invited to revise and resubmit their projects as an unsolicited proposal. Of those that we surveyed, only 1 of 8 respondents who had a proposal declined had resubmitted the proposal, while another respondent said that it was not made clear that proposals could be resubmitted.

Coordination of Transportation Requirements

Under the cooperative agreement, NASA is required to provide the ISS National Laboratory research facilities and resources and coordinate with CASIS when preparing CASIS-sponsored research for launch. CASIS has an integral role in the payload development and integration process during three distinct phases—pre-flight, operations, and post-flight. During the pre-flight phase, the CASIS operations team works with the researcher and implementation partners to understand project objectives and requirements such as power, crew and hardware compatibility needs, flight integration time frames, and design and integration support. CASIS submits the science objectives, requirements, and a development schedule to NASA. The NASA ISS National Laboratory Office also assigns staff to each CASIS-sponsored researcher to help coordinate and navigate the payload development and integration process and ensure that flight planning remains on track. During the operations and in-flight phase, CASIS provides operational support by collaborating with the implementation partner or the researcher to oversee

NASA's integration of the research project or hardware into the flight vehicle. The post-flight phase involves the return of payload samples or hardware from the flight vehicle to the researcher to begin post-processing activities, which CASIS monitors.

The ability to secure transportation for selected research investigations to the ISS facility is outside of CASIS's control and has presented challenges. NASA provides launch services to the ISS National Laboratory through its commercial resupply services contracts and CASIS receives cargo allocations for its sponsored research. Launch failures and delays, however, have resulted in cost increases. For example, the recent rocket launch failure to the ISS in October 2014 resulted in the loss of several CASIS-sponsored research investigations at a total cost of almost $175,000 which includes hardware and materials, labor consulting and grants. In addition, launch delays for another cargo resupply mission resulted in over $300,000 in cost increases for several researchers for additional materials and samples. CASIS officials explained that the majority of cost increases are related to biological research, which represents approximately 50 percent of the CASIS-sponsored research. These biological payloads have a limited viability or very specific requirements associated with the timing of the payload flight and often require consumables such as gas, nutrients, and water that must be replenished when a launch is delayed. Absorbing the increased cost has been a challenge for CASIS, but it is addressing the increased costs of delays by asking researchers that have biological payloads to identify the impact and associated costs for launch delays in their budgets so it can plan for budget reserves, if necessary.

Cooperative Efforts and Partnerships

The cooperative agreement requires CASIS to manage planning and coordination of research activities for both ground and on-orbit execution. According to CASIS officials, CASIS is addressing this requirement by leveraging the resources of companies that provide hardware, technical expertise and ground support. Eleven implementation partners have received over $5.4 million in funding from CASIS or its sponsored researchers since the establishment of CASIS through September 2014 to provide hardware, flight integration, and ground services for 58 research investigations. CASIS officials reasoned that by leveraging existing companies that can provide specialized hardware and integration capabilities on an as required basis, CASIS can effectively manage the ISS National Laboratory without having to maintain all the requisite skills or capabilities within its organization. CASIS-sponsored researchers are encouraged to select an implementation partner during the proposal submission process from a list of preferred partners. CASIS officials said that they assembled this list of implementation partners beginning with companies that had relationships with NASA for ISS-related operations and expanded the list through its own business development operations. These partners can provide hardware, and technical services, and consultation to researchers that address the project's science requirements and research needs aboard the ISS National Laboratory. Although CASIS provides a list of implementation partners, the researchers are responsible for entering into formal business arrangements with these partners and including the costs of the implementation partner support in their proposed budget. CASIS officials noted the cost can vary based on the amount of involvement required by the implementation partners and can range from $50,000 to $300,000 per flight.

Scientific Outreach and Educational Initiatives

In accordance with the cooperative agreement, CASIS is building a geographic network to facilitate outreach initiatives and cultivate new partnerships and has implemented educational initiatives that provide opportunities for educators and students to learn about and have access to the ISS National Laboratory. Specifically,

- **Network outreach:** CASIS has organized its outreach to scientific and academic communities in seven geographical areas. These areas are supported by more than 30 CASIS employees and consultants and each area has a research emphasis. See figure 4 for the locations of CASIS's networks and the research emphasis for each.

The outreach efforts conducted through CASIS's networks are primarily relationship-based and focused on engaging financial support, forging long term partnerships, and ultimately generating potential research and technology development projects for flight on the ISS National Laboratory. According to CASIS officials, academic institutions, research-specific organizations, philanthropic entities, and industry partners that CASIS identified through this network can benefit from the CASIS-sponsored research and technology development aboard the ISS National Laboratory. For example, Boston was identified as one of the geographic areas because it has over 100 universities and over 300 biotech companies that can support the commercialization of life sciences research and CASIS's mission.

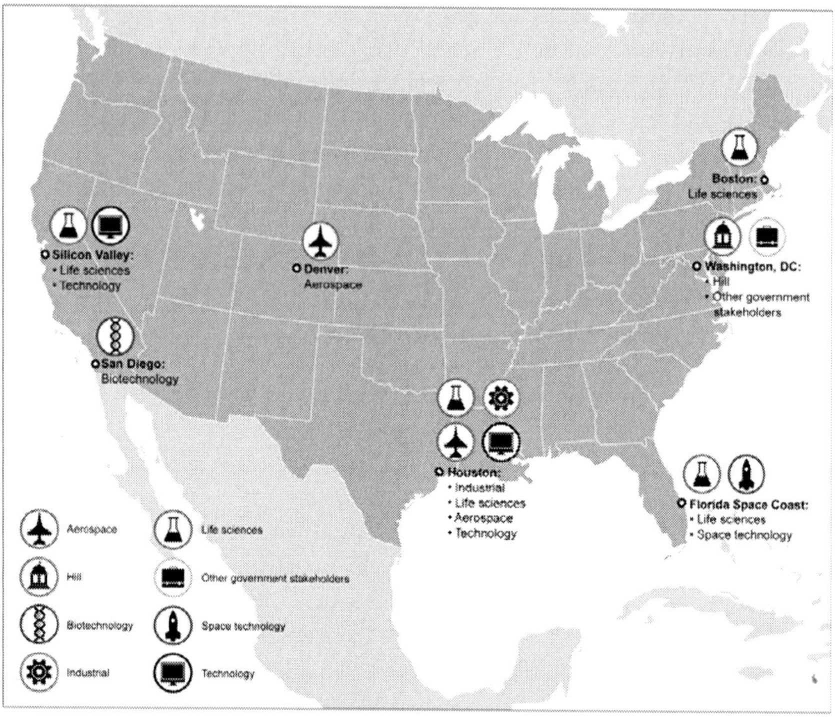

Source: GAO presentation of CASIS data. | GAO-15-397.

Figure 4. The Center for the Advancement of Science in Space (CASIS) Networks and Research Emphasis.

CASIS is working to expand its network. CASIS has developed 45 new partnerships to date and is leveraging a variety of new partnership opportunities. For example, in 2014, CASIS initiated two strategic campaigns, Good Earth—an international collaboration seeking to maximize ISS Earth observation capabilities—and Good Health—an effort to capitalize on the unique benefits of the microgravity environment so interventions can be developed to preserve health on Earth. CASIS officials expect both campaigns to bring together large scale collaborations to stimulate ISS utilization over the coming years. CASIS also supported the Rice Business Competition by providing a $25,000 grant during 2014 to a startup company that showed the most promise for developing a technology or business that would benefit from access to the ISS National Laboratory. This partnership also gives CASIS access to many forum events and panels.

According to CASIS officials, it has been challenging to raise additional funding from external sources to supplement the amount of funding provided by NASA to support and sustain its operations because CASIS is a new non-profit entity. Although CASIS's business development team is actively identifying partnerships and funding opportunities with commercial and non-profit granting organizations, CASIS officials said that it takes time to identify, develop, and mature these partnerships. CASIS and NASA officials said that the value of doing research aboard the ISS National Laboratory has to be further demonstrated so commercial industries can be convinced it is worth the high investment. Both NASA and CASIS officials said that demonstrating the value of research on the ISS as a substitute for ground-based research is a tremendous and important effort that is necessary to open a marketplace for space research. NASA officials stated that doing research aboard the ISS National Laboratory can take upwards of 2 to 3 years to plan and execute, time lines that are generally not acceptable to commercial companies that desire a more rapid return on their investments.

Ten of the 14 respondents to our survey reported that CASIS was effective in reaching out to the research community. For example, several researchers were made aware of CASIS opportunities by attending presentations from CASIS staff at industry meetings or campus visits. Respondents also offered areas for improvement for CASIS to increase utilization of the ISS National Laboratory. For example, five respondents said that CASIS could increase its visibility by attending more conferences, using more print ads, and working more with NASA on joint RFPs.

- **Education:** CASIS established its education strategic plan, which included building education programs that promote the ISS as a science, technology, engineering, and mathematics (STEM) learning platform; partnering with existing education entities such as schools, universities, and other educational foundations and associations; and reaching out to underrepresented and nontraditional demographics. CASIS also implemented various educational initiatives that it developed both internally and externally in conjunction with its partners. For example, in fiscal year 2014, CASIS supported 12 educational initiatives. CASIS sponsored the Space Station Academy, a 4-week online program designed to take participants on a simulated mission to the ISS as "virtual astronauts." This pilot program involved 25 students and 25 educators. In addition, CASIS supports its educational efforts through education grant funding and partnerships.

See appendix III for more information on additional CASIS educational initiatives.

NASA Has Not Staffed Advisory Committee with Which CASIS Is Required to Interact

The one required activity in the cooperative agreement that CASIS has been unable to address is its interaction with the ISS National Laboratory Advisory Committee (INLAC) because the committee has not been staffed by NASA. The NASA Authorization Act of 2008 required NASA to establish the INLAC under the Federal Advisory Committee Act. The INLAC was required to include membership from organizations that have formal agreements with NASA to utilize the U.S. portion of the ISS. As outlined in the 2008 act, this committee is required to exist for the lifespan of the ISS and is to function in an advisory capacity to the NASA Administrator by assessing and monitoring ISS National Laboratory resource utilization and reporting its assessments and recommendations at least annually.[12] According to the cooperative agreement, CASIS will coordinate with the INLAC as established under section 602 of the NASA Authorization Act of 2008 and review recommendations provided by the INLAC. Although NASA formally established the committee in 2009, NASA has not fully implemented the 2008 act because the committee has yet to be staffed. NASA officials told us that with CASIS in place, the great majority of non-NASA ISS users do not have an agreement with NASA because they work with CASIS. They added that there are exceptions where NASA works with other agencies, but those are typically for exploration technology or defense-related projects. In addition, NASA officials indicate that the INLAC has not been staffed because they believe that the structure and function of the current CASIS Board of Directors has proven to be a better alternative to a NASA advisory committee since the CASIS board represents a broad experience base including military, medical research, strategic partnerships, and engineering, among others. Further, NASA officials said that the Research Subcommittee of the Human Exploration and Operations Committee to the NASA Advisory Council also provides research advisory oversight of the ISS National Laboratory. This subcommittee's objectives, however, have a focus on human spaceflight and the membership of this subcommittee is to consist of individuals from the research committee with a broad awareness of human spaceflight related activities.

CASIS officials also believe that their board is performing some of the INLAC's advisory duties, but acknowledge that the board does not meet the section 602 requirements under the 2008 act—to monitor and report annually to the NASA Administrator its assessments and recommendations of ISS National Laboratory utilization—nor does its membership meet the criteria specified in the act. Without a staffed INLAC, NASA currently lacks a single advisory committee that represents all users of the ISS National Laboratory and provides ongoing monitoring and assessments and recommendations of ISS National Laboratory resource utilization, as required by the charter. As a result, CASIS is not able to fulfill its responsibilities as outlined in the cooperative agreement and as established under section 602 of the NASA Authorization Act of 2008.

CASIS and NASA Have Taken Steps to Measure and Assess CASIS Performance, but Measurable Targets Needed

CASIS's Metrics Consistent with Most Key Attributes, but Lack Quantifiable Goals

CASIS has established metrics, but not targets against which its performance can be measured by NASA. The metrics CASIS developed in collaboration with NASA for fiscal year 2015 meet most key attributes of successful performance measures. These metrics are based on CASIS responsibilities outlined in the cooperative agreement and are related to CASIS strategic goals and objectives. Metrics are included in an Annual Program Plan, which CASIS prepares with input from NASA.

We have previously reported that successful performance measures as a whole should have four general characteristics: demonstrate results, be limited to a vital few, cover multiple priorities, and provide useful information for decision making.[13] We cited specific attributes as key to successful performance measures, such as linkage, clarity, measurable targets, objectivity, and balance. The four characteristics are overarching, thus they do not necessarily directly link to the attributes. Furthermore, the attributes may not be equal, and a noted weakness does not mean that a measure is not useful. Weaknesses identified should be considered areas for further refinement. Table 3 defines the key attributes of successful performance measures.

Table 3. GAO's Key Attributes of Successful Performance Measures

Attribute	Definition
Linkage	Measure is aligned with goals and mission and clearly communicated throughout organization.
Clarity	Measure is clearly stated and the name and definition are consistent with the methodology used to calculate it.
Measurable Target	Measure has a numerical goal.
Objectivity	Measure is reasonably free from significant bias or manipulation.
Core Program Activities	Measures cover the activities that an entity is expected to perform to support the intent of the program.
Limited Overlap	Measure should provide new information beyond that provided by other measures.
Balance	Measures ensure that an organization's various priorities are covered.

Source: GAO. | GAO-15-397

We assessed CASIS's fiscal year 2015 metrics, and found that the metrics met almost all of these key attributes. The results of our assessment are shown in table 4. We also assessed the metrics CASIS had developed for fiscal year 2014, and similarly found that the metrics met most of the key attributes. The results of our assessment of CASIS's fiscal year 2014 metrics for key attributes of successful performance measures can be found in appendix IV.

Our analyses indicated that CASIS did not establish measurable targets or goals for either fiscal year 2014 or 2015 metrics, which limits its ability to use these metrics to assess performance. We have previously reported that performance metrics should have quantifiable, numerical targets or other measurable values, which help assess whether overall goals and objectives were achieved.[14] Without defined measurable targets or goals, it is unclear how NASA objectively assesses CASIS's performance.

CASIS officials noted that operating as a new entity with no history made it difficult to establish performance targets, but this is beginning to change. CASIS officials initially told us in July 2014 that establishing targets would be arbitrary because CASIS processes and metrics are still evolving. Subsequently, in January 2015, they indicated that since CASIS now has some operating history, they will be able to do so. The Chairman of the CASIS Board of Directors told us that measurable targets should be developed and that this is a priority for the Board.

Table 4. Assessment of the Center for the Advancement of Science in Space (CASIS) Fiscal Year 2015 Metrics Against GAO's Key Attributes of Successful Performance Measures

CASIS's Performance Metrics	Linkage	Clarity	Measurable Target	Objectivity	Core Program Activity	Limited Overlap	Balance
(1) CASIS grant calls issued	✓	✓		✓	✓	✓	✓
(2) Project proposals received from grant calls	✓	✓		✓	✓	✓	✓
(3) Project proposals awarded	✓	✓		✓	✓	✓	✓
(4) Project proposals received (non-CASIS grant)	✓	✓		✓	✓	✓	✓
(5) Project proposals awarded (non-CASIS grant)	✓	✓		✓	✓	✓	✓
(6) CASIS funding provided to projects	✓	✓		✓	✓	✓	✓
(7) Non-CASIS funding provided to projects	✓	✓		✓	✓	✓	✓
(8) Financial resources generated from non-NASA sources	✓	✓		✓	✓	✓	✓
(9) Direct, indirect, grant spending	✓	✓		✓	✓	✓	✓
(10) Flight projects manifested	✓	✓		✓	✓	✓	✓
(11) Flight projects delivered to ISS National Laboratory	✓	✓		✓	✓	✓	✓
(12) Allocation utilization[a]	✓	✓		✓	✓	✓	✓

Table 4. (Continued)

CASIS's Performance Metrics	Linkage	Clarity	Measurable Target	Objectivity	Core Program Activity	Limited Overlap	Balance
(13) Science, technology, engineering, and mathematics (STEM) projects executed	✓	✓		✓	✓	✓	✓
(14) Total reach of STEM projects	✓			✓	✓		✓
(15) Publications, patents, products resulting from flight projects	✓	✓		✓	✓	✓	✓
(16) CASIS outreach events	✓	✓		✓	✓	✓	✓
(17) Total media impact	✓			✓	✓		✓

Source: GAO analysis. | GAO-15-397

[a]Allocation utilization refers to utilization of the ISS National Laboratory and the information for this metric is provided by NASA.

However, CASIS has not established a date by which measurable targets will be developed. Further, CASIS officials indicated that not all metrics will have measurable targets initially because some metrics are subjective, such as those that attempt to measure the quality of research or a new technology generated by CASIS-sponsored research. The Chairman said that the CASIS Board of Directors is also working to develop targets for subjective measures, and they hope to have them in place in the next several years.

Although the ability to objectively measure performance is limited without measurable targets, CASIS and NASA officials generally agreed about how long-term success for CASIS will be defined. According to CASIS officials, success would ultimately be defined by demonstrating that the research and technology development performed aboard ISS National Laboratory benefits Earth and that commercial markets can be sustained in low-Earth orbit. NASA officials similarly said that developing commercial markets in space and bringing products back to Earth will determine success.

NASA Assesses CASIS Annually, but Performance Assessment Is Not Documented

NASA performs an annual assessment of CASIS's performance consistent with its responsibilities in the cooperative agreement, but this assessment is not documented.[15] The Cooperative Agreement Technical Officer (CATO) uses the metrics in CASIS's quarterly and annual reports to monitor CASIS's efforts.[16] The cooperative agreement also requires CASIS to propose an adjustment to the metrics if performance is not going to be met. However, without performance targets, CASIS cannot determine whether the metrics need to be adjusted. Further, without these targets, NASA and CASIS cannot conduct assessments that are measurable or conclusive and, therefore, the assessments are subjective. According to the CATO, during the annual program review, he assesses CASIS metrics for trends, looking for improvements over time and questioning any perceived lack of progress. The CATO added that he discusses any issues identified during the annual review with CASIS officials, NASA management, and stakeholders. CASIS officials concurred, and told us this discussion with NASA highlights areas for further refinement. For example, as a result of such discussion, CASIS is now more proactively engaging NASA technical expertise on available flight hardware, and has broadened business development efforts aimed at attracting new commercial users of the ISS National Laboratory.

Both CASIS and NASA officials told us that NASA does not document its annual program review of CASIS performance. Federal standards for internal controls call for information to be recorded and communicated to management and others who need it to carry out their responsibilities.[17] This type of documented information is important to support decision making and conduct assessments. CASIS officials have not asked for a formal summary of the results of NASA's annual program review because CASIS receives informal feedback on quarterly reports provided to NASA. CASIS also maintains minutes of regularly scheduled meetings with NASA where any issues that need to be discussed between CASIS and NASA are addressed. While NASA does not document this annual assessment, NASA officials told us that they were generally satisfied with CASIS performance. CASIS officials, however, said that the results of the annual review should be reported in some sort of formal manner to make the information more actionable.

Because CASIS is allocated at least 50 percent of ISS research capacity, future success of the ISS as a research platform is partially dependent on the efforts CASIS has undertaken. However, without definitive and documented assessment factors, NASA will be challenged to take action in response to CASIS performance. For example, without documentation, NASA lacks support to terminate the cooperative agreement, if deemed necessary. Conversely, NASA also would have no record to justify extending the cooperative agreement to support a possible ISS life extension. The cooperative agreement will expire at the end of fiscal year 2020, but includes a provision for an extension.[18]

CONCLUSION

The ISS offers the potential for scientific breakthroughs, a unique test bed for new technologies and applications, and a platform for increased commercial and academic research. Achieving greater utilization of the ISS and its unique capabilities, showing the benefit of commercial and academic research, and demonstrating success to generate increased interest from potential users could help NASA get a better return on its significant investment in the ISS. NASA currently lacks an advisory committee established under the Federal Advisory Committee Act that is composed of individuals representing organizations who have formal agreements with NASA to use the U.S. portion of the ISS. As a result, CASIS is not able to fulfill its responsibility as outlined in the cooperative agreement that requires it to coordinate with INLAC as established under the NASA Authorization Act of 2008 and review recommendations originated by the INLAC. A fully staffed and operational INLAC could provide information to senior NASA management on how to better utilize the constrained resources of the ISS—which could affect how CASIS attracts new users and fulfills its responsibility to increase utilization of the ISS National Laboratory.

In addition, clearly defined measurable targets are essential for CASIS to demonstrate results, allow NASA to objectively assess CASIS performance, and help stakeholders assess whether overall goals and objectives for the ISS National Laboratory are achieved. Finally, NASA's annual performance assessment of CASIS is not documented and the results are provided to CASIS on an informal basis. Not documenting the results of the annual program assessment is a practice contrary to good internal controls, which call for information to be recorded and communicated to management and others who need it to carry out their responsibilities, to include taking appropriate corrective actions. Without a clear, well-documented assessment of CASIS performance, NASA management and stakeholders could also be missing information important for decision making, for example, deciding to extend the cooperative agreement with CASIS beyond the September 2020 expiration if the service life of the ISS is extended or terminate the agreement, if necessary.

RECOMMENDATIONS FOR EXECUTIVE ACTION

We recommend that the NASA Administrator take the following three actions:

- In order for NASA to fully implement the NASA Authorization Act of 2008 and for CASIS to fulfill its responsibility as outlined in the cooperative agreement, direct the Associate Administrator for the Human Exploration and Operations Mission Directorate to fully staff the INLAC.
- In order to set clear goals to allow NASA to objectively assess CASIS performance, require the ISS Program Manager work with CASIS to collectively develop and approve measurable targets for metrics for fiscal year 2016 and beyond.
- In order to provide CASIS management actionable information to better fulfill its responsibilities and NASA management with

additional information by which to make future decisions concerning the extension of the agreement with CASIS, require the ISS Program Manager to document the annual program assessment of CASIS performance.

AGENCY AND THIRD-PARTY COMMENTS AND OUR EVALUATION

NASA and CASIS each provided written comments on a draft of this report. NASA and CASIS also provided technical comments, which have been incorporated into the report, as appropriate. NASA partially concurred and CASIS non-concurred with one of our recommendations and both NASA and CASIS concurred with the other two recommendations.

NASA partially concurred and CASIS non-concurred with our recommendation directing the Associate Administrator of the Human Exploration and Operations Mission Directorate within NASA to staff the INLAC. In response to this recommendation, both NASA and CASIS raised concerns that the current requirements for membership of the INLAC would create a conflict of interest. Specifically, NASA stated that the individuals who would make up the committee would likely have user agreements with CASIS and, in many cases, would be receiving funding from CASIS and NASA. Furthermore, because these entities would be competing for CASIS resource allocations, CASIS believes that they would not be sufficiently independent to perform the functions required of the committee. In response to these concerns, CASIS indicated the composition of membership as defined in the NASA Authorization Act of 2008 should be amended. NASA also responded that while meeting statutory obligations and obtaining knowledgeable input and recommendation to achieve optimal utilization of the ISS is important, it is the agency's position that the CASIS Board of Directors serves the intent of the INLAC charter by providing recommendations regarding effective utilization of the ISS. As a result, NASA indicated that it plans to work with the Congress to adjust the INLAC requirement to address these concerns. We continue to believe our recommendation is valid. We do not see that staffing the INLAC as directed in the 2008 act would necessarily result in a conflict of interest and that the entities would be competing for CASIS resource allocations. The act required an advisory committee that represents all users of ISS National Laboratory and that provides ongoing monitoring and assessment and makes recommendations. According to the cooperative agreement between CASIS and NASA, CASIS is directed to coordinate with the INLAC and review the committee's recommendations. The INLAC, however, functions only in an advisory capacity; therefore we do not see how a conflict of interest would be created by the membership of the INLAC. Furthermore, according to the NASA Authorization Act of 2010, CASIS shall be guaranteed access to not less than 50 percent of the United States research capacity allocation. Because CASIS has to agree with NASA for an allocation of resources at a level below 50 percent, we do not see how the composition of the INLAC would create a competition for resource allocation with CASIS. In addition, it was not clear to us in our review that the existing mechanisms in place accomplish these requirements. If NASA were to seek relief or changes to this requirement, it should clearly outline how these requirements can be met through existing bodies and processes.

NASA and CASIS concurred with our recommendation directing the ISS Program Manager to work with CASIS to collectively develop and approve measurable targets for metrics in fiscal year 2016 and beyond. In response to this recommendation, NASA stated that fiscal year 2016 is a reasonable time to establish measurable targets with CASIS because the non-profit will be entering its fourth full year of operations. Similarly, CASIS responded that it is now in a position to develop targets for key metrics and plans to formalize the process in fiscal year 2016. NASA indicated that these targets should be established by December 31, 2015. Once complete, this action should address our recommendation to develop and approve measurable targets for CASIS's metrics.

NASA and CASIS also concurred with our recommendation directing the ISS Program Manager to document the annual program assessment of CASIS performance. In response to this recommendation, NASA said that it would begin documenting the agency's annual program assessment in response to CASIS's 2015 annual report. Once complete, this action should address our recommendation to document NASA's annual assessment of CASIS's performance.

Sincerely yours,
Marie A. Mak Director
Acquisition and Sourcing Management

APPENDIX I: OBJECTIVES, SCOPE, AND METHODOLOGY

Our objectives were to assess the extent to which (1) the Center for the Advancement of Science in Space (CASIS) has initiated and implemented the required management activities for research aboard the International Space Station (ISS) National Laboratory, and (2) the National Aeronautics and Space Administration (NASA) and CASIS measure and assess CASIS's performance.

To determine the extent to which CASIS is performing the required management activities for non-NASA research aboard the ISS National Laboratory, we obtained and reviewed relevant legislation and documentation, and interviewed ISS program and CASIS officials. We reviewed the NASA Authorization Act of 2005, which designated the U.S. Operating Segment of the ISS as a National Laboratory; the NASA Authorization Act of 2008, which directed NASA to establish an ISS National Laboratory Advisory Committee; and the NASA Authorization Act of 2010, which required NASA to enter into a cooperative agreement with a nonprofit organization to manage the activities of the ISS National Laboratory. We also reviewed the cooperative agreement between NASA and CASIS and the CASIS fiscal year 2014 and 2015 Annual Program Plans for CASIS responsibilities related to the required activities outlined in Section 504(c) of the NASA Authorization Act of 2010. We examined the CASIS portfolio management and research prioritization process and various market analyses and studies that CASIS considered in establishing research areas. We reviewed the CASIS proposal review and evaluation process for solicited and unsolicited proposals as well as the Requests for Proposals that CASIS had issued to solicit research proposals. We studied fiscal year 2014 quarterly and annual reports to gain insight into the activities CASIS had undertaken to meet its responsibilities. We reviewed CASIS business development efforts, including funding and marketing processes and outreach efforts. We reviewed the partnerships CASIS has established with philanthropic institutions that could provide additional resources to sponsor research aboard the ISS National laboratory and implementation partners that provide logistical assistance to researchers. Additionally, we reviewed CASIS education efforts, particularly science, technology, engineering, and mathematics activities. We also reviewed GAO, NASA Inspector General, and NASA reports on sustaining the ISS. We interviewed several ISS program officials including the ISS Program Director, ISS Program Manager, ISS Program Scientist, and the Cooperative Agreement Technical Officer to gain their perspectives on the work CASIS was performing. We also interviewed officials in the Space Life and Physical Sciences Research and Applications division, including the NASA Liaison to CASIS, to gain perspective on the work NASA is sponsoring aboard the ISS. In addition, we interviewed the CASIS President and Executive Director, the CASIS Chief Operating Officer, the CASIS Chief Financial Officer, and the Chairman of the CASIS Board of Directors to better understand the processes and procedures being implemented, how proposals are evaluated, and the challenges that CASIS faces to further implement the responsibilities outlined in the cooperative agreement.

To obtain additional information on CASIS's performance and the effectiveness of its implementation of some of the required activities, we used information provided to us by CASIS to select a random sample of 20 principal investigators who had submitted either a solicited or unsolicited research proposal to CASIS. Of the 20 researchers selected, we conducted structured interviews with 14 researchers to obtain additional insights into CASIS's performance. Although the randomly selected researchers are, in part, representative of the population of 172 researchers who had submitted proposals to CASIS through July 2014, the descriptive nature of the responses and the relatively small sample size does not permit the development of reliable, quantitative estimates that are generalizable to the population. However, we believe our interview results provide us with valuable information about researcher's experiences and perspectives on CASIS's performance in the area of soliciting, reviewing and providing feedback on proposals.

To determine whether CASIS, in collaboration with NASA, has established performance metrics, we reviewed CASIS metrics as presented in its fiscal years 2013 to 2015 Annual Program Plans. We concentrated on fiscal year 2014 and 2015 metrics, but examined the previous metrics to determine how performance measures evolved. We also reviewed CASIS quarterly reports for fiscal year 2014 and the first quarter of fiscal year 2015 and the fiscal year 2014 annual report to determine how performance was measured and reported to NASA. We analyzed CASIS's fiscal year 2014 and 2015 metrics to evaluate whether they adhered to GAO's key attributes of successful performance measures, which were identified in previous work. Judgment was required to determine which attributes were applicable to assess and whether the performance measures met the definition of the attributes selected. To determine how NASA assesses CASIS performance, we reviewed the cooperative agreement to determine relevant NASA responsibilities, including the roles of the Cooperative Agreement Technical Officer and NASA Liaison. We also interviewed the NASA Liaison to CASIS, the

Cooperative Agreement Technical Officer, and CASIS officials to gain their perspective on the evolution of metrics and how they are used to assess CASIS's performance.

Our work was performed at NASA Headquarters in Washington, D.C., and Johnson Space Center in Houston, Texas. We also visited CASIS headquarters in Melbourne, Florida.

We conducted our review from April 2014 to April 2015 in accordance with generally accepted government auditing standards. These standards require that we plan and perform the audit to obtain sufficient, appropriate evidence to provide a reasonable basis for our findings and conclusions based on our audit objectives. We believe that the evidence obtained provides a reasonable basis for our findings based on our audit objectives.

APPENDIX II: INTERNATIONAL SPACE STATION NATIONAL LABORATORY MANAGEMENT ACTIVITIES REQUIRED IN THE NASA AUTHORIZATION ACT OF 2010

The National Aeronautics and Space Administration (NASA) Authorization Act of 2010 directed that the Administrator shall provide initial financial assistance to the organization with which the Administrator enters into a cooperative agreement to manage the International Space Station (ISS) National Laboratory.[1] In August 2011, after a competitive process, NASA signed a cooperative agreement with the Center for the Advancement of Science in Space, Inc. (CASIS), a not-for-profit entity, to manage the activities of the ISS National Laboratory through September 30, 2020. The 2010 act outlined several management and research and development activities for CASIS, as the not-for-profit entity selected, to initiate, as follows:

1) Planning and coordination of the ISS national laboratory research activities.
2) Development and implementation of guidelines, selection criteria, and flight support requirements for non-NASA scientific utilization of ISS research capabilities and facilities available in United States-owned modules of the ISS or in partner-owned facilities of the ISS allocated to United States utilization by international agreement.
3) Interaction with and integration of the International Space Station National Laboratory Advisory Committee established under section

602 of the National Aeronautics and Space Administration Authorization Act of 2008 (42 U.S.C. 17752) with the governance of the organization, and review recommendations provided by that Committee regarding agreements with non-NASA departments and agencies of the United States Government, academic institutions and consortia, and commercial entities leading to the utilization of the ISS national laboratory facilities.

4) Coordination of transportation requirements in support of the ISS national laboratory research and development objectives, including provision for delivery of instruments, logistics support, and related experiment materials, and provision for return to Earth of collected samples, materials, and scientific instruments in need of replacement or upgrade.
5) Cooperation with NASA, other departments and agencies of the United States Government, the States, and commercial entities in ensuring the enhancement and sustained operations of non-exploration-related research payload ground support facilities for the ISS, including the Space Life Sciences Laboratory, the Space Station Processing Facility and Payload Operations Integration Center.
6) Development and implementation of scientific outreach and education activities designed to ensure effective utilization of ISS research capabilities including the conduct of scientific assemblies, conferences, and other fora for the presentation of research findings, methods, and mechanisms for the dissemination of non-restricted research findings and the development of educational programs, course supplements, interaction with educational programs at all grade levels, including student focused research opportunities for conduct of research in the ISS national laboratory facilities.
7) Such other matters relating to the utilization of the ISS national laboratory facilities for research and development as the Administrator may consider appropriate.

Appendix III: The Center for the Advancement of Science in Space (CASIS) Educational Activities

Educational Initiatives	Objectives	Status
The Center for the Advancement of Science in Space (CASIS) Developed Educational Initiatives		
National Design Challenge Pilot Project	A national education campaign that provides educators and their students the opportunity to design and implement an authentic research experiment on the International Space Station (ISS).	Houston, Tex. – Six educators and 220 students completed experiments to fly to the ISS on Orb-3 in October 2014. Denver, Colo. – Three schools are currently developing experiments that will be sent to the ISS in spring on 2015. The pilot includes 105 middle and high school students.
CASIS Academy Live	Brings middle and high school students to the Kennedy Space Center Visitor Complex and the Space Life Science Lab to interact with an astronaut and research scientist to send their experiment to the ISS.	Six CASIS Academy Live events have been held at the Space Life Sciences Lab and the Kennedy Space Center for 390 Central Florida middle and high school students.
CASIS Academy Website	Created to educate middle school students about the ISS.	There have been nearly 15,000 total views of the CASIS Academy student website with themonthly average of 2,492 views. The educators' webpage has a total of 1,530 views averaging 255 monthly.

Educational Initiatives	Objectives	Status
CASIS Fellows	Work with volunteers across the nation who communicate the CASIS mission and information about recent research conducted on board the ISS National Laboratory. Volunteers serve as pilot tester, focus group and provide local training on CASIS education programs.	Program brought awareness of the ISS and CASIS science, technology, engineering, andmathematics (STEM) activities to 450 educators and students at various workshopsand presentations.
PGA Stem Golf Camp	Partnership with the Professional Golfers' Association of America Center for Golf Learning and Performance, Cobra Puma Golf and St Lucie County Schools to bring together science and golf by offering a 5-day golf summer cap to underprivileged middle school students teaching them math and physics .	Sixty-three middle school students participated in the Professional Golfers' Association of America STEM Camp in summer 2014.
CASIS Educational Partnership Programs		
Zero Robotics Middle School Program	Five-week summer program for middle school students to work in teams with program staff, mentors, and scientists to learn about programming, robotics and space engineering while getting hands-on experience working with and programming Synchronized Position Hold, Engage, Reorient, Experimental Satellites.	There were 550 students and 110 teachers from 9 different states who participated in the program in summer 2014.
BioServe Ants in Space	Offers students the ability to participate in near real-time life science research onboard the ISS to study foraging ant behavior.	The ant experiment was flown to the ISS in December of 2013. A total of 8,814 students in 32 states participated in program in FY2014.

Appendix III. (Continued)

Educational Initiatives	Objectives	Status
Story Time From Space	Videotapes of astronauts reading selected stories from the ISS.	The videotapes were downloaded in January 2014. Fundraising efforts continue for Phase 2 in parallel with the development of the demonstration kit of materials that will complement the science content in the books.There have been 6,500 student s and educators participating in the program in 2014.
Student Spaceflight Experiments Program	Students engage in the experiments design and proposal writing process that culminates in flying an experiment on the ISS.	CASIS is a national sponsor of Missions 5 and 6 in fiscal year 2014. This represents more than 8,000 students actively engaged in authentic research experiences. CASIS presented to 400 of these students and their parents at the Student Spaceflight Experiments Program National conference in Washington D.C.
Space Station Academy	A 4-week online program designed to take participants on a simulated mission to the ISS as "virtual astronauts". Offered to middle and high school students and children and adults outside of the school system.	A total of 25 students and 25 educators participated in the prototype version of the Space Station Academy in July 2014.
NASA HUNCH (High School Student United with NASA to Create Hardware)	The High School Students United with NASA to Create Hardware program is a partnership between high schools and NASA where students design, build and implement an experiment in microgravity.	The experiment is being developed by a teamof students at Lakewood High School in Colorado.

Educational Initiatives	Objectives	Status
National Geographic Learning	CASIS entered into a partnership with National Geographic Learning/Cengage to help develop an online interactive science program for grades K-6.	

Source: GAO presentation of CASIS data. | GAO-15-397.

APPENDIX IV: ASSESSMENT OF THE (CASIS) FY 2014 METRICS AGAINST GAO'S KEY ATTRIBUTES OF SUCCESSFUL PERFORMANCE MEASURES

Performance Metric	Linkage	Clarity	Measurable Target	Objectivity	Core Program Activity	Limited Overlap	Balance
(1) Description of business development plan to proactively generate interest and market opportunities as well as stimulate unsolicited proposals	✓			✓	✓		
(2) Number and description of research pathways identified by CASIS that will be targeted via grants and business development	✓	✓		✓	✓	✓	✓
(3) Number of grants issued by CASIS, by research pathway	✓	✓		✓	✓	✓	✓
(4) Number of responsive proposals meeting evaluation criteria, per grant	✓			✓	✓	✓	✓

Appendix IV. (Continued)

Performance Metric	Linkage	Clarity	Measurable Target	Objectivity	Core Program Activity	Limited Overlap	Balance
(5) Number of unsolicited proposals received by CASIS, by research pathway and meeting evaluation criteria	✓			✓	✓	✓	✓
(6) Number of unsolicited proposals received by CASIS, meeting evaluation criteria, and not within targeted research pathways	✓			✓	✓	✓	✓
(7) Summary reports of disciplines responding, by grant and unsolicited proposal submission	✓	✓		✓	✓	✓	✓
(8) Percentage of all proposals from single and multi-disciplinary teams				✓	✓	✓	✓
(9) Number of awards given to solicited proposals In each research pathway	✓	✓		✓	✓		✓
(10) Dollar ($) amount of awards given to solicited proposals in each research pathway	✓	✓		✓	✓	✓	✓

Performance Metric	Linkage	Clarity	Measurable Target	Objectivity	Core Program Activity	Limited Overlap	Balance
(11)Dollar ($) amount of NASA funding allocated to grants vs. direct vs. indirect costs	✓	✓		✓	✓	✓	✓
(12) Dollar ($) amount of non-NASA funding allocated to grants	✓	✓		✓	✓		✓
(13) Breakdown of direct, indirect, grants	✓	✓		✓	✓		✓
(14) Total amount of government vs. non-government funding comprising CASIS budget/financial posture	✓	✓		✓	✓	✓	✓
(15) Actual vs. projected expenditures	✓	✓		✓	✓	✓	✓
(16) Report economic impacts from use of the ISS National Laboratory, if available	✓			✓	✓	✓	✓
(17) Number of flight projects manifested	✓			✓	✓		✓
(18) Number of total flight projects manifested as a result of solicited proposals or investments	✓	✓		✓	✓		✓

Appendix IV. (Continued)

Performance Metric	Linkage	Clarity	Measurable Target	Objectivity	Core Program Activity	Limited Overlap	Balance
(19) Number of total flight projects manifested as a result of unsolicited proposals or investments	✓			✓	✓		✓
(20) Describe intended impacts/ outcomes of ISS NL research and development to life on Earth	✓			✓	✓	✓	✓
(21) Report scientific or technological breakthroughs related to use of the ISS NL	✓			✓	✓		✓
(22) Report transformational/ translational science	✓			✓	✓		✓
(23) Report projects or activities contributing to national scientific, educational, or technology initiatives	✓			✓	✓	✓	✓
(24) Number of proposers, by solicited/unsolicited, and by business development activity	✓			✓	✓	✓	✓
(25) Number of proposals from new partnerships (formed to engage with CASIS/ISS)	✓	✓		✓	✓		✓

Performance Metric	Linkage	Clarity	Measurable Target	Objectivity	Core Program Activity	Limited Overlap	Balance
(26) Report new initiatives to solicit interest in/engagement with CASIS toward broader utilization of the ISS	✓	✓		✓	✓	✓	✓
(27) Number of awards given to unsolicited proposals	✓	✓		✓	✓	✓	✓
(28) Dollar ($) amount given to unsolicited proposals	✓	✓		✓	✓	✓	✓
(29) Number and dollar ($) amount of awards by type of responding organization (other government agencies, academic, individual, commercial, other)	✓	✓		✓	✓		✓
(30) Dollar ($) amount contributed to projects by non-CASIS sources, and their origins (including targeted giving, commercial entities, private investments)	✓	✓		✓	✓		✓
(31) Dollar ($) amount and description of flight projects provided by other government agencies	✓	✓		✓	✓		✓

Appendix IV. (Continued)

Performance Metric	Linkage	Clarity	Measurable Target	Objectivity	Core Program Activity	Limited Overlap	Balance
(32) Describe actual impacts of ISS NL research and development to life on Earth (specific examples, as they occur)	✓	✓		✓	✓	✓	✓
(33) Number of projects resulting in publication or application, as they occur	✓	✓		✓	✓		✓
(34) Number of publications from projects related to CASIS activities, in all areas	✓	✓		✓	✓		✓
(35) Number of patents from projects related to CASIS activities, in all areas	✓	✓		✓	✓	✓	✓
(36) Number of products or technologies currently in development on government funding, related to CASIS activities ("pipeline")	✓	✓		✓	✓	✓	✓
(37) Number of products or technologies in development on non-government funding, related to CASIS activities	✓	✓		✓	✓	✓	✓

Performance Metric	Linkage	Clarity	Measurable Target	Objectivity	Core Program Activity	Limited Overlap	Balance
(38) Report number and market value (if available) of products brought to market related to CASIS activities	✓	✓		✓	✓		✓
(39) Government contributions to CASIS, including like-kind, philanthropic giving, commercial investment, institutional investment, funded partnerships, etc.	✓			✓	✓		✓
(40) Report change (improvements) in awareness among key stakeholder groups	✓			✓	✓	✓	✓
(41) Report change (improvements) in requests for information from CASIS from responsible parties	✓			✓	✓	✓	✓

Source: GAO analysis. | GAO-15-397

End Notes

[1] National Aeronautics and Space Administration Authorization Act of 2005, Pub. L. No.109-155, § 507 and National Aeronautics and Space Administration Authorization Act of 2010, Pub. L. No. 111-267, § 504.

[2] In September 2014, NASA awarded contracts to The Boeing Company and Space Exploration Technologies Corporation to develop a commercial crew capability to transport astronauts to and from the ISS. NASA expects this capability to be available in fiscal year 2018. Currently, NASA purchases seats—at a cost of over $65 million each in 2015—aboard Russian Soyuz space capsules to transport crew to and from the ISS.

[3] Pub. L. No. 109-155, § 507.

[4] National Aeronautics and Space Administration Authorization Act of 2008, Pub. L. No. 110-422, § 602.

[5] Pub. L. No. 111-267, § 504.

[6] A completed description of the activities can be found in appendix II.

[7] The Administration recently proposed extending the operational life of the ISS from 2020 to at least 2024. Prior GAO work has shown that it is technically feasible to extend the ISS operational life to at least 2028.

[8] Federal Acquisition Regulation (FAR) § 35.003.

[9] Pub. L. No.111-267, § 504(c).

[10] We did not include the last activity outlined in the 2010 act—related to addressing other matters pertaining to the utilization of the ISS National Laboratory for research and development as the Administrator may consider appropriate—since no specific CASIS responsibilities were delineated in the Cooperative Agreement.

[11] Accelerators and competitions represent grants awarded through CASIS partnership with third parties that help to accelerate business development. For example, CASIS uses the Mass Challenge to look at entities that might have a business model suitable for developing new or improving existing materials for the ISS. Mass Challenge will help the companies establish their businesses, and will also provide financial support to those start-ups in the top tier of the challenge. The 2014 Mass Challenge program resulted in a review of 128 finalists. CASIS worked with these finalists to prepare proposals which were submitted for review. CASIS partnered with Boeing to award three projects for a total flight prize of $600,000.

[12] Pub. L. No. 110-422, § 602, codified at 51 U.S.C. § 70906.

[13] GAO, *Tax Administration: IRS Needs to Further Refine Its Tax Filing Season Performance Measures*, GAO-03-143 (Washington, D.C.: Nov. 22, 2002). The characteristics are based on Government Performance and Results Act of 1993 (GPRA), Pub. L. No. 103-62, as modified by GPRA Modernization Act of 2010, Pub. L. No. 111- 352, guidance on assessing strategic planning, performance measurement, and reporting in the federal government. GPRA was enacted, in part, to hold federal agencies accountable for achieving program results.

[14] GAO-03-143.

[15] The Cooperative Agreement, section 4.3, also requires CASIS to comply with the requirements of OMB Circular A-110.

[16] The Cooperative Agreement Technical Officer, a NASA employee within the ISS Program Office at Johnson Space Center, is charged with oversight of the cooperative agreement.

[17] GAO, *Standards for Internal Control in the Federal Government*, GAO/AIMD-00-21.3.1 (Washington, D.C.: Nov. 1, 1999).

[18] NASA Grant and Cooperative Agreement, section 4.19 Multiple Year Grant or Cooperative Agreement.

End Note for Appendix I

[1] Pub. L. No. 111-267, § 504.

In: The International Space Station
Editor: Evelyn Clemens

ISBN: 978-1-63484-071-2

Chapter 2

HEARING CHARTER FOR "THE INTERNATIONAL SPACE STATION: ADDRESSING OPERATIONAL CHALLENGES"*

House Committee on Science, Space and Technology; Subcommittee on Space OR Committee on Science, Space, and Technology; Subcommittee on Space

PURPOSE

At 9:00 a.m. on Friday, July 10, 2015, the Subcommittee on Space will hold a hearing titled *The International Space Station: Addressing Operational Challenges*. The purpose of this hearing is to examine the current status of the International Space Station (ISS). The Subcommittee will evaluate the National Aeronautics and Space Administration's (NASA) plans for dealing with operational and maintenance challenges, the status of the ISS partnership, how NASA is utilizing the ISS to enable future deep space exploration, and the Administration's request to extend ISS operations to 2024.

---

* This is an edited, reformatted and augmented version of a document presented July 10, 2015 for a hearing conducted by the House Committee on Science, Space, and Technology, Subcommittee on Space.

BACKGROUND

The ISS is one of the most complex and expensive man-made structures ever built.[1] The ISS is a joint project among five participating space agencies—NASA, Roscosmos (Russian Space Agency), the European Space Agency (ESA), Japanese Aerospace Exploration Agency (JAXA), and Canadian Space Agency (CSA). The ISS orbits approximately 250 miles above the Earth's surface once every 90 minutes at five miles per second. Weighing in at nearly one million pounds, it is the length of a football field (including end zones) and has the equivalent working and living space of a six bedroom home. The pressurized living space is approximately equal to a Boeing 747. The solar arrays used to power the Station would cover nearly an acre and eight miles of electrical wiring powers various sections of the facility. Out of the global space launch market, launches to the ISS accounts for approximately 15 percent of the total. For 2013-2014, this was 25 out of 168 total launches worldwide.[2]

Among other benefits, the ISS provides a proving ground for NASA's human exploration technologies and other NASA mission directorates and various federal agencies as well as a microgravity laboratory for private companies. The ISS generally operates with a rotating crew of six astronauts from the U.S. and international partner space agencies, though at times this drops to three crewmembers as is the case at present.

The United States currently launches cargo resupply missions to the ISS through the Commercial Resupply Services (CRS) contract. The Russian *Progress* and Japanese *HTV* also provide cargo resupply to the ISS.

The two U.S. providers under this contract, Orbital ATK and Space Exploration Technologies Corporation (SpaceX), provide cargo delivery services on a firm fixed-price contract. The two providers carry supplies for the astronauts as well as science and research equipment to the ISS in conjunction with international partners.

NASA currently lacks a domestic capability to ferry astronauts to and from the ISS. American astronauts (and international partners) are reliant on the Russian Soyuz which launches from the Baikonur Cosmodrome in Kazakhstan. NASA's existing contract for Russian Soyuz runs through 2017 and costs roughly $75 million a seat.[3] NASA recently announced that it is negotiating a new contract with Russia for services from 2017 to 2018.[4] Last fall, NASA signed contracts with Space Exploration Technologies (SpaceX) and the Boeing Company to develop and provide transportation to and from the ISS through the Commercial Crew Program (CCP).

Once the Commercial Crew program begins ferrying astronauts to the ISS, NASA may be able to add a seventh crew member. This has the potential of increasing research and utilization time on the station equivalent to approximately 35 hours a week.[5] In fiscal year 2014 (FY14), the Station hosted 368 experiments (a 28 percent increasc from FY13). These included 64 in biology and biotechnology, 91 in Earth and space science, 50 educational activities, 36 in human research, 43 in physical science, and 84 in technology.[6] The utilization of the unique microgravity environment on the ISS is augmented by the Center for Advancement of Science in Space (CASIS). This non-profit organization was chosen in 2011 in accordance with the NASA Authorization Act of 2010.[7]

Budget

Budget Authority ($ in millions)	Actual 2014	Enacted 2015	Request 2016	FY15 vs FY16	Notional			
					2017	2018	2019	2020
Space Operations	3,774.0	3,827.8	4,003.7	175.9	4,191.2	4,504.9	4,670.8	4,864.3
International Space Station	2,964.1	-	3,105.6	-	3,273.9	3,641.0	3,826.0	4,038.3
Space and Flight Support (SFS)	809.9	-	898.1	-	917.3	863.8	844.8	826.1

The prime contractor for operations and maintenance of the ISS is the Boeing Company. This accounts for $1.106 billion of the overall Space Operations account in the FY16 budget request. Operations and maintenance includes managing resources, logistics, systems, and operational procedures. Additionally, the operations and maintenance project manages resource requirements and changes, including vehicle traffic, cargo logistics, stowage, and crew time. The project is also responsible for providing anomaly resolution and failure investigations as needed.

Commercial Cargo

The Commercial Spaceflight program at NASA began in 2006 by funding multiple companies to develop systems for transporting cargo to the ISS with an eye towards eventually having multiple carriers compete for the resupply contract. This was accomplished through the Commercial Orbital Transportation Services (COTS) and Cargo Resupply Services (CRS) programs. At this point, both of the companies involved, Orbital ATK and SpaceX, have successfully delivered cargo to the ISS. While the SpaceX contract includes a down-mass capability (returns cargo to Earth), Orbital ATK's *Cygnus* spacecraft (like the Russian Progress, European Space Agency's ATV or the Japanese Space Agency's HTV) has no down-mass capability. In 2008, NASA signed two CRS contracts. The SpaceX contract is valued at $1.6 billion for 12 missions and Orbital ATK contract is valued at $1.9 billion for eight missions.[12] Regardless of anomalies or accidents, the cost associated with these launches is set and paid out in increments as various milestones are met through the manufacturing process. The final payment is not made unless the payload docks with the ISS and delivers the cargo. In some cases, substantial portions of the contract may be paid out prior to delivery of the cargo.[13]

On October 28, 2014, Orbital ATK attempted to launch its *Cygnus* cargo ship to the ISS carrying 5,000lbs. of supplies and science experiments. Approximately 15 seconds after launch the Antares rocket suffered a catastrophic failure and the rocket as well as the *Cygnus* were lost. The investigation into the exact cause is ongoing. In the meantime, Orbital ATK procured the services of United Launch Alliance (ULA) to launch its next *Cygnus* payload to the ISS. This flight is required under the company's CRS contract and it is tentatively scheduled for launch in October. As the CRS contract is firm-fixed price, the use of a different rocket and altered launch conditions will not financially impact the government.

On April 28, 2015, the Russian *Progress* resupply vehicle, M-27M launched from Kazakhstan carrying 5,196lbs of cargo to the ISS. After achieving orbit, the *Progress* vehicle suffered several anomalies that resulted in the loss of the vehicle. The Progress-60 launched on July 3, 2015 carrying 6,100lbs of cargo and docked with the ISS on July 5, 2015.

On June 10, 2015, while performing various software testing procedures there was an inadvertent firing of thrusters used to control the orbit of the ISS. The Russian and American engineering teams are the process of identifying the root cause, but as of yet have not released any public findings.

On June 28, 2015, SpaceX attempted to launch its seventh cargo resupply mission to the ISS. At approximately 139 seconds after lift-off, the rocket suffered a catastrophic failure that resulted in the loss of the vehicle. The unmanned *Dragon* capsule was carrying approximately 5,000 pounds of pressurized cargo, including research experiments, food, crew provisions and exercise equipment. Additionally, the vehicle was carrying a replacement spacesuit for EVA activity, this suit was necessary after water started seeping into the helmet of one of the spacesuits in December 2013. The cause of the SpaceX launch failure is still under investigation. As with the Orbital ATK launch failure, the contractor will be responsible for the investigation with participation and oversight from NASA, Federal Aviation Administration (FAA), and the National Transportation Safety Board (NTSB) in accordance with FAA regulations for licensed commercial launches.

The extent to which these three failures over the course of eight months will affect the ISS program is unclear. NASA announced that the astronauts aboard the Station are in no immediate danger and that they have enough food, water, and oxygen to last until September. The next resupply mission of a Japanese HTV is scheduled to be launched in August.[14]

Commercial Crew

NASA awarded fixed-price contracts to Boeing and SpaceX in September 2014 for the Commercial Crew program. The total potential values of these contracts are $2.6 billion for SpaceX and $4.2 billion for Boeing for a total of $6.8 billion over the life of the contracts. These two companies will proceed through the final design, development, testing, evaluation and human rating certifications under a fixed-price contract.

The President's budget request for FY16 includes $1.24 billion for the Commercial Crew Program. This would be an increase of 54 percent over the appropriated funding for FY2015 ($805 million). In testimony before this Committee earlier this year, a NASA official stated that

> "[i]f NASA does not receive the full requested funding for CCtCap in FY 2016 and beyond, NASA will have to adjust (delay) milestones for both partners proportionally and extend sole reliance on Russia for crew access to the ISS. The partners may request contract cost adjustments and the certification dates will be delayed."

Federal Acquisition Regulations (FAR)[15] and the commercial crew contracts[16] do allow NASA to adapt their acquisition strategy for one or two contractors to accommodate varying appropriation levels. Despite the $6.8 billion projected value of the contracts, NASA has never completed an independent cost estimate of the Commercial Crew Development Program or the program estimates that the companies provided for their funding requirements;[17] however, the contracts are fixed-price, meaning they are capped at the agreed upon levels. This does not guarantee that the contractors will not need to be bailed-out in the event that they are unable to complete the contractual work.

The NASA Authorization Act of 2010 authorized $312 million, $500 million, and $500 million for the Commercial Crew Program for fiscal years 2011, 2012 and 2013 respectively. NASA has consistently requested more funding for Commercial Crew than the program has been authorized or previously appropriated.[18] Three years ago, the NASA Administrator testified before the Committee that the FY13 request would put NASA "on track" for a commercial crew capability by 2017.[19] The actual appropriation for FY13 was $305 million less than the request. Two years ago, the Administrator testified to the Committee that NASA was still on track for a 2017 launch date, but full funding of the FY14 request was "essential" to enabling Commercial Crew access to the International Space Station by 2017.[20] The actual appropriation for FY14 was $125 million less than the request. The FY16 NASA budget justification states that 2017 is still the target date for a Commercial Crew capability. The Committee-passed NASA Authorization Act for FY16 and FY17 included full funding for the Commercial Crew program. Funding history for the program is included below.

Funding History *$ in millions*	**Program Phase**					
Partner	CCDev1	CCDev2	CCiCap	CPC1	CCtCap	Total
Paragon	1.40	-	-	-	-	1.40
United Launch Alliance	6.70	-	-	-	-	6.70
Blue Origin	3.70	22.00	-	-	-	25.70
Sierra Nevada	20.00	105.60	227.50	10.00	-	363.10
SpaceX	-	75.00	460.00	9.60	2,600.00*	3,144.60
Boeing	18.00	112.90	480.00	9.90	4,200.00*	4,820.80
Total Funding	49.80	315.50	1,167.50	29.50	6,800.00	8,362.30

Source - http://www.nasa.gov/sites/default/files/files/CCtCapFactSheet.pdf.

* Represents total potential value of the contract.

Orion As a Backup

In addition to the use of commercial crew contractors, NASA is required under federal law to ensure that the *Orion* crew vehicle has the capability to dock to the ISS in an emergency.[21] The law,[22] which was passed by Congress and signed by the President, was very specific in describing this as a "minimum capability requirement"[23] for the capsule. It is important to note that the law does not require *Orion* to be launched by the Space Launch System (SLS). While *Orion* may not be an efficient vehicle to conduct ISS transportation missions, the ability to conduct a mission to the ISS would provide redundancy and additional options for access. At present, NASA is not building the *Orion* with the capability to service the ISS. In an interview with *Space News* in June 2014, Administrator Bolden expressed his view of the legal requirement for *Orion* to serve as a backup capability:

> "It's a bad, bad day when you have to send Orion to the international space station because it means either we've lost each of the [commercial] vehicles that was designed to do that through some accident, or they failed or something. So we don't want to have to rely on Orion to do that.
>
> "We made a commitment to industry we would not compete with them.
>
> "If we had said, 'We're going to keep Orion as a backup,' there were serious doubts as to whether industry would have made the investment at all in a commercial crew vehicle because their assumption was, 'OK, if NASA is going to build a vehicle to go to low Earth orbit, what is NASA going to want to use?' Naturally, they're going to want to use their own vehicle.
>
> "So Orion, while it probably can — or will — be capable of going to the international space station, is not designed to do that, is not intended to do that."[24]

Given the subsequent launch failures with the Orbital ATK and SpaceX cargo vehicles as well as the loss of the Russian *Progress* vehicle over the course of the past year, the potential loss of commercial crew capabilities seem to have a higher likelihood of possibly occurring.

It is unclear how NASA would handle such a situation given the current state of the development efforts of *Orion*. The Administration has consistently requested less than has been previously appropriated for the program. In the FY13, FY14, and FY15 budget requests, the Administration asked for reductions of $175.1 million, $87 million, and $144.2 million from previous year funding.[25] It is difficult to assess the ability of NASA to press the *Orion* into service for emergency crew transfer capabilities while simultaneously requesting reductions to the budget for the program.

ISS Utilization

Since inception of the ISS, utilization of the facility has been an ongoing concern of Congress. There are several factors that limit the amount of research that can be conducted on the ISS at any given time; the most limiting of these is crew time to perform the experiments. Additionally, NASA's Inspector General reported last year that only about 41 percent of crew time was used on utilization and research.[26] NASA claims that, once the Commercial Crew contractors are providing consistent service, the addition of a seventh crew member could nearly double research time.

The NASA Authorization Act of 2010 required NASA to procure the services of a non-profit entity to run the National Laboratory portion of the ISS. NASA awarded this cooperative agreement to the Center for the Advancement of Science in Space (CASIS). The GAO recently reviewed the progress CASIS has made towards satisfying the requirements of their agreement. Generally speaking, the organization is making good progress, but there are still questions about the efficacy of the National Lab. According to GAO it is difficult to assess the progress made by CASIS in fulfilling its requirements under the cooperative agreement because NASA has not established a formal process for doing so:

> "Using the established metrics, NASA is required by the cooperative agreement to perform an annual program review of CASIS's performance. This review is informal and not documented as ISS program officials provide the results to CASIS orally. This approach is inconsistent with federal internal control standards, which call for information to be recorded and communicated to those who need it to manage programs, including monitoring performance and supporting future decision making. Although NASA officials reported that they were generally satisfied with CASIS's performance, CASIS officials said a formal summary of the results would make the information more actionable."[27]

In addition to CASIS activities, NASA is required to maximize use of the ISS. The NASA Authorization Act of 2010 requires NASA to "sustain the capability for long-duration presence in low-Earth orbit, initially through continuation of the ISS and full utilization of the United States segment of the ISS as a National Laboratory, and through assisting and enabling an expanded commercial presence in, and access to, low-Earth orbit, as elements of a low-Earth orbit infrastructure."[28] Additionally, NASA is required to utilize the ISS as a "testbed" for technologies developed for future human exploration in deep space.[29] Finally, NASA must "maximize the productivity and use of the ISS with respect to scientific and technological research and development, advancement of space exploration, and international collaboration."[30]

For fiscal year 2016, NASA requested $394 million for the purpose of microgravity research aboard the ISS. This represents approximately 9.8 percent of the total budget for the Space Operations account or about 35 percent of what it costs for the operations and maintenance of the program. The ISS currently supports activities for the Science Mission Directorate space and Earth science payloads, Space Technology Mission Directorate, Human Research Program (HRP), and Advanced Exploration Systems (AES). In addition to the ISS research program, NASA utilizes the ISS as a testbed for technology development in the Space Technology Mission Directorate and the Science Mission Directorate. The NASA Authorization Act of 2015 passed by the House by unanimous consent explicitly authorized the use of the ISS for this purpose.[31]

The ongoing experiments, future experiments, and ISS utilization efforts can be reviewed in detail at http://www.nasa.gov/mission pages/station/main/index.html. The most notable of these experiments include:[32]

- Study on the effects of long-term spaceflight on astronaut Scott Kelly and cosmonaut Mikhail Kornienko. Both astronauts will spend almost a full year in space.
- Leveraging microgravity environment for protein crystal grown and disease models to aid study of human diseases such as Huntington's, cystic fibrosis, ALS, and others.
- Creation of a searchable publicly accessible database on biological flight data for use by academia, industry, and other federal agencies to retrieve and analyze science conducted on organisms flown in space.
- Multi-generational, long duration fruit fly laboratory research to aid in human research. As fruit flies share 77% of human disease genes, this provides for a unique study scenario.
- Development of Cold Atom Laboratory with research teams that includes three Nobel Laureates.

- Demonstrate 3D in-space printing as a first step in the "machine shop" capability for long-duration deep space human exploration.
- Conduct Bigelow Expandable Activity Module (BEAM) demonstration. This activity supports the AES program in understanding inflatable habitat use in deep space.
- Continuing operation of the Alpha Magnetic Spectrometer (AMS), a state of the art particle physics detector with a primary mission to find evidence of dark matter. The initial positron science results have been referenced in 279 astrophysics/physics publications.

While NASA's use of the ISS has improved in recent years, there are still areas of concern with regards to utilization. There is a lack of a unified federal government strategy for utilization of the facility. The NASA Authorization Act of 2015 requires the Administration to develop a federal government-wide utilization plan.[33] Additionally, the recent launch accidents resulted in the loss of research equipment and technology assets that may limit the utilization of the Station for at least the immediate future.

ISS Extension to 2024

Last year, the Administration proposed extending the life of the ISS from 2020 to 2024, but has not provided its plan to get commitments from the international partner space agencies or budget requirements for the extension.[34] NASA is currently authorized to operate the ISS until 2020 under current law.[35] The extension of the program requires legislative action. As the Administration works to get the international partner nations to agree to extend the ISS, the House-passed NASA Authorization Act of 2015 includes a requirement that NASA provide an extension criteria report to Congress. This report would provide Congress with a baseline and notional plan to evaluate the merits of extending the program.[36]

The NASA Inspector General (IG) released a report on ISS extension in September 2014. The IG found several areas of concern with regards to extension. First, that NASA had not identified major structural obstacles to extension but that several risk areas required mitigation. According to the report,

> "First, the ISS faces a risk of insufficient power generation due in part to faster than expected degradation of its solar arrays. Second, although most replacement parts have proven more reliable than expected, sudden failures of key hardware have occurred requiring unplanned space walks to repair or replace hardware. Third, although NASA has a robust cargo transportation system, it has a limited capacity to transport large replacement parts – such as solar arrays and radiators – to the Station."[37]

The IG also found that cost projections for extension appeared overly optimistic. According to NASA officials, the budget for the ISS will remain between \$3 billion and \$4 billion annually through 2024. In the judgment of the IG, "this estimate is based on overly optimistic assumptions and the cost to NASA will likely be higher."[38] NASA does not have a public estimate on the costs associated with extension. However, a recent Congressional Budget Office (CBO) cost estimate on the ISS extension provision in S. 1297, The U.S. Commercial Space Launch Competitiveness Act,[39] estimates that, should appropriations be allocated under the extension, it would cost the government approximately \$14.3 billion over the next ten years to extend the ISS operating life to 2024.[40]

Key Questions

1. What are the costs associated with extension of the ISS and how are those costs measured against the benefits of extension?
2. How can Congress assess whether the ISS program is meeting its goals and objectives and how can those metrics inform the costs and benefits of extension?
3. If Congress does not extend ISS operations beyond 2020, what impact would that have on the U.S. Space Program?
4. What can NASA expect to gain by extending the ISS beyond 2024?
5. Is there an optimal date for extension beyond 2020?
6. Are there any technical concerns that would limit extension beyond a certain date?
7. How does NASA plan to mitigate the extension concerns expressed by the Inspector General?

8. What are the impacts on utilization of the ISS associated with the three cargo flight failures and how will those impacts be mitigated by NASA and the international partners?
9. Will ISS crew be limited as a result of the cargo failures, and if so, how will that impact utilization?
10. What steps have been taken to mitigated risks associated with reliance on private contractors for crew and cargo transportation?
11. Has NASA taken the appropriate steps to ensure that the *Orion* can serve as a backup to the commercial crew contractors in an emergency situation?

End Notes

[1] The NASA Inspector General estimates that the United States has invested almost $75 billion which includes "$43.7 billion for construction and program costs through 2013, plus $30.7 billion for 37 supporting Space Shuttle flights, the last of which took place in July 2011." Extending the Operational Life of the International Space Station Until 2024 (IG-14-031). Retrieved at: https://oig.nasa.gov/audits/reports/FY14/IG-14-031.pdf.

[2] Launch history is based on data compiled by the FAA's Office of Commercial Space Transportation (AST) Annual Commercial Space Transportation Compendium found at: https://www.faa.gov/about/office org/headquarters offices/ast/media/FAA Annual Compendium 2014.pdf.

[3] Ibid. 1, p. 21.

[4] NASA Sole-source Procurement Announcement of Crew Transportation and Rescue Services from Roscosmos https://prod.nais.nasa.gov/cgibin/eps/synopsis.cgi?acqid=163919.

[5] ISS utilization projections are courtesy of NASA delivered via staff briefings in February and March of 2015 in support of the President's Budget Request.

[6] President's Budget Request for Fiscal Year 2016 for the National Aeronautics and Space Administration, Congressional Justification. P. SO-14. Retrieved at http://www.nasa.gov/sites/default/files/files/FY2016_Budget_Book_508_TAGGED.pdf.

[7] Section 504 of the National Aeronautics and Space Administration Authorization Act of 2010. Retrieved at https://www.congress.gov/111/plaws/publ267/PLAW-111publ267.pdf.

[8] Title 3, H.R. 2578, Commerce, Justice, Science and related agencies Appropriations Act of 2016. Retrieved at https://www.congress.gov/bill/114th-congress/house-bill/2578/text/rh?q={%22search%22:[%22%22hr2578%22%22]}.

[9] Based on estimates from NASA's FY15 spending plan submitted to Congress

[10] Title 3, H.R. 2578, S. Rep. 114-66. Retrieved at https://www.congress.gov/114/crpt/srpt66/CRPT-114srpt66.pdf.

[11] Ibid. 9.

[12] NASA Awards Space Station Commercial Resupply Services Contracts. Retrieved at http://www.nasa.gov/home/hqnews/2008/dec/HQ_C08-069_ISS_Resupply.html.

[13] Commercial Cargo: NASA Management of Commercial Orbital Transportation Services and ISS Commercial Resupply Contracts. NASA Office of Inspector General released on June 13, 2013. Retrieved at https://oig.nasa.gov/audits/reports/FY13/IG-13-016.pdf.

[14] Statement by NASA Administrator Charles F. Bolden, Jr. June 29, 2015. Retrieved at https://www.nasa.gov/press-release/nasaadministrator-statement-on-the-loss-of-spacex-crs-7

[15] Federal Acquisition Regulations 52.249-2, Termination for Convenience of the Government (Fixed-Price). Retrieved at http://www.ecfr.gov/cgi-bin/text-idx?SID=d3e3b99553af16 abc0188213eee065c3&mc=true&node=se48.2.52_1249_62&rgn=div8

[16] Clause B.4 Post Certification Missions, Clause H.4 NFS 1852.232-77 Limitation of Funds (fixed-price contract), Clause H.8 Post Certification Mission Task Ordering Procedures (Applicable to CLIN 002), Commercial Crew Transportation Capability (CCtCAP) Contract with Boeing (NNK14MA75C), Commercial Crew Transportation Capability (CCtCAP) Contract with SpaceX (NNK14MA74C). Retrieved at http://www.nasa.gov/centers/ kennedy/about/foia/reading room.html#.VZKzV lVhBc.

[17] NASA contracted with Booz|Allen|Hamilton to complete an independent cost assessment of the program which was released on March 1, 2013 and can be found here http://www.nasa. gov/pdf/741617main_CCP-ICA-DRD-2e-Public-Releaseable-FinalReport-3-5-13-508.pdf. However, as noted by the NASA Inspector General, "the assessment found that the estimates were optimistic, and that the Program was likely to experience cost growth. In addition, Booz Allen noted that without costs projected over the life of the Program, NASA officials will not be able to independently evaluate each partner's progress."

[18] FY2011 request: $500 million. FY2011 actual: $307 million. FY2012 request: $850 million. FY2012 actual: $392 million. FY2013 request: $830 million. FY2013 actual: $525 million. FY2014 request: $821 million. FY2014 actual: $696 million.

[19] Charles F. Bolden, Jr., Administrator, National Aeronautics and Space Administration, statement before the House Committee on Science, Space, and Technology, March 7, 2012.

[20] Charles F. Bolden, Jr., Administrator, National Aeronautics and Space Administration, statement before the House Committee on Science, Technology, and Space, Subcommittee on Space, April 24, 2013.

[21] Title 42, §18323(b)(3), United States Code.

[22] National Aeronautics and Space Administration Authorization Act of 2010 (P.L. 111-267). Retrieved at http://www.gpo.gov/fdsys/pkg/PLAW-111publ267/pdf/PLAW-111pub l267. pdf.

[23] Title 42, §18323(b), United States Code.

[24] Klotz, Irene, "Orion No Backup for Commercial Crew, Says Bolden," SpaceNews, June 18, 2014. Accessed at http://www.spacenews.com/article/civil-space/40957orion-no-backup-for-commercial-crew-says-bolden.

[25] President's Budget Requests for Fiscal Year 2013, Fiscal Year 2014, and Fiscal Year 2015.

[26] Ibid. 1, p. 7.

[27] Government Accountability Office report #GAO-15-397, "Measurable Performance Targets and Documentation Needed to Better Assess Management of National Laboratory." Retrieved at http://www.gao.gov/assets/670/669851.pdf.

[28] Section 202, National Aeronautics and Space Administration Authorization Act of 2010 (P.L. 111-267). Retrieved at http://www.gpo.gov/fdsys/pkg/PLAW-111publ267/pdf/PLAW-111publ267.pdf.

[29] Section 308(c), National Aeronautics and Space Administration Authorization Act of 2010 (P.L. 111-267). Retrieved at http://www.gpo.gov/fdsys/pkg/PLAW-111publ267/pdf/PL AW-111publ267.pdf.

[30] Section 502(a) National Aeronautics and Space Administration Authorization Act of 2010 (P.L. 111-267). Retrieved at http://www.gpo.gov/fdsys/pkg/PLAW-111publ267/pdf/PLAW-111publ267.pdf.

[31] Section 213 and Section 503, H.R. 810, National Aeronautics and Space Administration Authorization Act of 2015, retrieved at http://www.gpo.gov/fdsys/pkg/BILLS-114hr810rfs/pdf/BILLS-114hr810rfs.pdf.

[32] Examples of ISS experiments and utilization are courtesy of NASA via staff briefings delivered in February and March of 2015 in support of the President's Budget Request.

[33] Section 211(g), H.R. 810, National Aeronautics and Space Administration Authorization Act of 2015, retrieved at http://www.gpo.gov/fdsys/pkg/BILLS-114hr810rfs/pdf/BILLS-114hr810rfs.pdf.

[34] Statement by NASA Administrator Charles F. Bolden, Jr. January 8, 2014 http://blogs.nasa.gov/bolden/2014/01/08/obamaadministration-extends-international-space-station-until-at-least-2024/.

[35] Title 51, §70907, United States Code.

[36] Section 211(f) National Aeronautics and Space Administration Authorization Act of 2015, retrieved at http://www.gpo.gov/fdsys/pkg/BILLS-114hr810rfs/pdf/BILLS-114hr810rfs.pdf.

[37] Ibid. 1, p. ii.

[38] Ibid. 1, p. iii.

[39] Section 13, S. 1297, U.S. Commercial Space Launch Competitiveness Act. Retrieved at http://www.gpo.gov/fdsys/pkg/BILLS114s1297is/pdf/BILLS-114s1297is.pdf.

[40] Congressional Budget Office Cost Estimate of S. 1297. Retrieved at http://www.cbo.gov/sites/default/files/114th-congress2015-2016/costestimate/s1297.pdf.

In: The International Space Station
Editor: Evelyn Clemens
ISBN: 978-1-63484-071-2

Chapter 3

STATEMENT OF WILLIAM H. GERSTENMAIER, ASSOCIATE ADMINISTRATOR FOR HUMAN EXPLORATION AND OPERATIONS, NASA. HEARING ON "THE INTERNATIONAL SPACE STATION: ADDRESSING OPERATIONAL CHALLENGES"*

Mr. Chairman and Members of the Subcommittee, thank you for the opportunity to appear before you today to discuss the status of the International Space Station (ISS) Program, as well as our efforts to conduct microgravity research and promote U.S. economic activity in low-Earth orbit (LEO), and expand human spaceflight beyond LEO. The ISS represents an unparalleled capability in human spaceflight that is increasing our knowledge of basic physics and biology. This knowledge is benefiting our lives here on Earth and enhancing the competitiveness of private industry here in the United States. The research and technology demonstrations onboard the ISS are providing the basis for extending human presence beyond the bounds of LEO and taking our next steps into the proving ground of cis-lunar space. With the Administration's proposed extension of the ISS until at least 2024, NASA has the opportunity to more fully utilize research, and commercial and

* This is an edited, reformatted and augmented version of a statement presented July 10, 2015 before the House Committee on Science, Space, and Technology, Subcommittee on Space.

international partnerships to ensure that the U.S. continues to be the world leader in human spaceflight and to enable U.S. industry to realize the commercial benefits of research and development in the microgravity environment of space.

The ISS is vital to NASA's mission to extend human presence into the solar system. In order to prepare for human expeditions into deep space, we must first use the unique environment of ISS to conduct the research and technology demonstrations necessary to keep our crews safe and productive on long-duration spaceflights. The ISS – which has been home to a continuous human presence on orbit for almost 15 years – is NASA's only long-duration flight analog for future human deep-space missions, and it provides an invaluable laboratory for research with direct application to the exploration requirements that address human risks associated with deep-space missions. It is the only space-based multinational research and technology test bed available to identify and quantify risks to human health and performance, identify and validate potential risk mitigation techniques, and develop countermeasures for future human exploration. As NASA learns more about the changes to the human body from spaceflight and develops countermeasures to support long-duration missions, this same research is providing unique insight into problems facing our aging terrestrial population. Through the National Laboratory and the Center for the Advancement of Science in Space (CASIS), U.S. companies are taking advantage of new research opportunities that may provide a competitive edge. Across a range of disciplines and applications, ISS research ultimately benefits people on Earth.

Cargo and Crew Transportation

In order to realize the full potential of the ISS' capabilities, the platform is serviced by a fleet of operational vehicles, including two U.S. cargo resupply vehicles: Space Exploration Technologies' (SpaceX) Dragon and Orbital ATK's Cygnus. These two providers have flown a combined total of eight cargo missions to the ISS under the Commercial Resupply Services (CRS) contracts, which were awarded on December 23, 2008. The cargo flights have demonstrated the viability of the Government use of commercially provided services – rather than owning and operating the spacecraft and launch vehicles – for the delivery of experiments, supplies, and spares to the Station, enabling NASA to focus its development efforts on deep-space vehicles to take our astronauts beyond LEO. Further, both companies have demonstrated their resourcefulness in dealing with challenges. NASA anticipates awarding one or more CRS-2 contracts later this year.

The overall cargo strategy of having multiple providers has served NASA well, and the importance of having multiple providers is critical for assured cargo access to the ISS, as has been demonstrated by recent losses of Orbital ATK Cygnus, Russian Progress, and SpaceX Dragon cargo flights to ISS. NASA was able to continue ISS operations and research by relying on SpaceX after the Cygnus cargo vehicle anomaly last fall. Orbital is expected to recover and return to flight before the end of 2015. Orbital ATK acquired a launch on an Atlas V vehicle while its Antares launcher is outfitted with a new engine. This strategy allows Orbital ATK to use its Cygnus vehicle and resume cargo delivery capability later this year. The ability of Orbital ATK to quickly integrate Cygnus with an Atlas V is a direct result of their experience in integrating their satellites onto different launch vehicles. The basic CRS-1 contract premise of obtaining cargo services enabled Orbital ATK to creatively acquire a new launch vehicle. This demonstrates the effectiveness of our current contract strategy. Late last month, the SpaceX-7 cargo mission experienced an anomaly during launch. SpaceX has demonstrated extraordinary capabilities in its first six cargo resupply missions to the Station, and we know they can replicate that success. We will work with and support SpaceX to assess what happened, understand the specifics of the failure and correct it to move forward. These events are reminders that spaceflight is difficult, but we learn from each success and each setback, and our commercial cargo program was designed to accommodate the loss of cargo vehicles. With the delivery last week of supplies aboard a Russian Progress vehicle, and with the prospect of a Japanese H-II Transfer Vehicle (HTV) flight in August, we do not anticipate any need to delay the upcoming Soyuz 43S launch later this month, which will return us to six-crew operations and research.

In the area of crew transportation, while NASA continues to develop Commercial Crew capability to provide crew transportation and rescue services to the ISS, the Russian Soyuz spacecraft currently provides these services. The ISS routinely hosts six crewmembers on long-duration missions with the support of two Soyuz spacecraft. The limit in crew size is driven by the Soyuz three-crew-carrying capability. There are currently four Soyuz missions per year to accomplish ISS crew rotations.

NASA's plans for research through 2017 are based having six crew on ISS, including the Human Research Program objectives we need to accomplish during this period to keep on track to reduce or retire risks for deep-space exploration. It should be noted, however, that Station has – from time to time – hosted only three crew during brief transition periods. Before the loss of the SpaceX-7 cargo flight, the current period of three-crew operations had resulted in the deferral of a docking adapter installation, as well as the deferral of some Node 1 preparation tasks. In addition, a rodent experiment was dropped from the SpaceX-7 cargo flight, and some fluid shift experiments were moved to the next Expedition. At this point, NASA anticipates that the current period of having three crew aboard will not last longer than a few weeks, after which, we will staff back up to six crew and resume the normal rate of research activities. Beyond 2017 when U.S. crew providers come on line, our plans count on having an ISS crew complement of seven. Even after losing the docking adapter on SpaceX-7, NASA will have time to add docking capability to ISS to support the first U.S. commercial crew flights.

In 2014, NASA contracted with two U.S. providers for crew transportation and rescue services for ISS. The Commercial Crew Transportation Capability (CCtCap) contracts will complete the development of domestic systems to provide safe, reliable, cost-effective access to and from ISS. SpaceX's Crew Dragon and Boeing's CST-100 spacecraft will begin ferrying our crews to Station from U.S. soil by the end of 2017, contingent upon receiving the full amount requested in the FY 2016 Budget Request, enhancing the robustness of our transportation system and ending our sole reliance on Russia for the provision of these services. U.S. commercial crew capabilities will enable the Station crew to be expanded from six to seven astronauts and cosmonauts, resulting in a doubling of on-orbit research time to almost 80 hours per week. This is because the seventh crew member will be able to focus his or her time almost exclusively on conducting experiments, rather than on Station operations and maintenance.

I want to thank this Committee for authorizing full funding in FY 2016 for our Commercial Crew Program. It is vitally important that NASA receive this funding level to keep the development of these systems on track for flights in 2017. If the Agency is funded with a Continuing Resolution for the first quarter of FY 2016, NASA will need to address how it will fund our partners' development activities at the current contractual schedule. The CCtCap contractors are only required to work on milestones to the extent that NASA has obligated funding for those milestones. If funding is not available in FY 2016 for the initial FY 2017 milestones, the contractors will have to stop work or work at risk until additional funding can be obligated, existing CCtCap contracts will need to be renegotiated, most likely resulting in schedule delays and increased contract cost, and NASA will need to continue to rely solely on Russian Soyuz capability to meet America's requirements for crew transportation services. NASA has no plans to downselect the number of partners in response to lower-than-requested funding levels. As experience has shown with cargo, NASA's plan to establish a redundant crew transportation capability is critically important for robust, safe ISS operations. This redundancy is even more critical during the development phase. We appreciate the Committee's support for our plan to end sole reliance on Russia for crew transportation through contracts with two U.S. providers.

With over 350 American companies across 36 states working toward this goal, there are significant economic benefits to returning these launches to American soil. At the same time, every dollar we send overseas rather than investing at home represents an investment we could be making in ourselves rather than in the Russian economy.

There are also longer term fiscal considerations to consider. NASA projects that the average seat price will be $58 million per seat for Commercial Crew. The currently contracted seat price for Soyuz for 2017 is approximately $76 million per seat.

Sustainability of ISS and Extension to 2024

The ISS continues to be a very healthy system operating well within prudent technical margins, and consistently demonstrating outstanding steady-state performance that meets or exceeds prior engineering estimates. While systems were originally specified to be both reliable and maintainable, the operational experience NASA and its Partners are gaining is providing invaluable information on reliability and maintainability standards for future application to spacecraft design and mission planning. This enables systems needed for long-duration spaceflight to be tested in preparation for missions far from Earth for which reliability and maintainability are absolutely required. Just as short-duration Space Shuttle flights prepared us for long duration Station flights, ISS is preparing us for missions that will not have the option of immediate crew return in the event of an anomaly.

In January 2014, the Administration announced its intent to extend ISS operations until at least 2024. The research we will conduct on ISS through 2024 will be essential to the safe and effective conduct of human exploration beyond LEO. This extension is also critical to commercial sector planning for the use of the ISS. Industry requires the planning stability provided by the extension in order to consider further investment in microgravity research and transportation services. Commercial LEO development, spurred in part by the continuation of ISS, will also help enable exploration and make NASA resources available for deeper space exploration.

In addition to the United States, the Government of Canada has announced that Canada will continue its participation in the ISS to 2024. There have been multiple public indications that Russia will continue participating in the ISS program through 2024; Roscosmos (the Russian Federal Space Agency) has publicly commented that it expects to receive government authority by the end of the year to continue ISS beyond 2020. The Government of Japan has also indicated that its decision to support ISS operations beyond 2020 will likely be made in the near future after internal government deliberations are completed. The European Space Agency is expected to address ISS operations and utilization beyond 2020 at their ministerial meeting in late 2016. The ISS Partners have expressed support for continuing research on ISS, and see tremendous benefit for extended research opportunities.

As NASA has moved into Station's intensive utilization phase, we have become more cost-efficient in ISS operations. In the FY 2016 President's Budget Request, ISS Operations and Maintenance (O&M) is only 35 percent of the ISS request. The majority of the request, 55 percent, is for ISS Crew and Cargo Transportation. The remaining 10 percent is for ISS Research. Since the ISS was extended to 2020 in 2011, NASA has reduced the ISS O&M budget through a combination of efficiencies in sustaining activities, some content reductions, and cutbacks in operations overhead. While NASA continues to look for further efficiencies, we have already achieved a level of efficiency that allows us to productively operate and sustain the ISS, keep our crews healthy and safe, and support utilization with substantially reduced resources. Ongoing activities to responsibly lower the O&M cost of the ISS include changes to our contracts to incentivize efficiency, lower overhead cost, and apply targeted enhancements in technology investments to reduce manpower-intensive processes. These activities are assumed in the FY 2016 President's Budget Request.

ISS Research

The ISS supports research across a diverse array of disciplines, including high-energy particle physics, Earth remote sensing and geophysics experiments, molecular and cellular biotechnology experiments, human physiology research (including bone and muscle research), radiation research, plant and cultivation experiments, combustion research, fluid research, materials science experiments, and biological investigations. In addition, the ISS is an invaluable platform for technology development efforts. Research and development conducted aboard the ISS holds thc promise of next-generation technologies, not only in areas directly related to NASA's exploration efforts, but in fields that have numerous terrestrial applications. The ISS will provide these opportunities to scientists, engineers, and technologists through at least 2024. Beyond being a feat of unparalleled engineering and construction, as well as international collaboration, the ISS is a place to learn how to live and work in space over a long period of time and foster new markets for commercial products and services. Remarkably, 83 countries/areas worldwide have participated in ISS utilization.

NASA's Human Research Program continues to develop biomedical science, technologies, countermeasures, diagnostics, and design tools to keep crews safe and productive on long-duration space missions. The progress in science and technology driven by this research could have broad impacts on Earth as it advances our ability to support long-duration human exploration.

On March 27, 2015, NASA astronaut Scott Kelly and cosmonaut Mikhail Kornienko of Roscosmos launched to the ISS to begin a one-year mission aboard the orbiting outpost. NASA and Roscosmos selected several collaborative investigations for this mission to evaluate the effects of long-duration spaceflight on humans. Each of the U.S. investigations will be grouped into one of seven categories: functional, behavioral health, visual impairment, metabolic, physical performance, microbial, and human factors. Researchers expect the mission's investigations to provide data on biomedical, performance, and behavioral changes and challenges astronauts may face when they embark on longer-duration missions, like those to an asteroid, Mars, or beyond. Data from the expedition will be used to determine whether there are ways to further reduce the risks on future long-duration missions to an asteroid and eventually Mars.

The investigations involving astronauts Scott and Mark Kelly, who are identical twins, will provide NASA and outside researchers with a genetic blueprint and broader insight into the subtle genetic effects and changes that may occur during long-term (i.e., one year) spaceflight as compared to Earth-based environments. The studies will focus on four areas: human physiology, behavioral health, microbiology/microbiome, and molecular or -omics studies (-omics refers to a system-level approach to studying molecular biology; examples include genomics, proteomics, and metabolomics). Although the investigations conducted on the Kelly brothers are not expected to provide definitive data about the effects of spaceflight on individuals — because there are only two subjects for data collection — they do serve as a demonstration project for future research initiatives. These investigations may identify changes to pursue in research of larger astronaut populations.

NASA is also exploring open-source science where databases are made available to a large number of researchers for investigation. This approach is in contrast to the past practice of one researcher "owning" all of the data from their investigation. This open-science approach shares a large data set of information with researchers for a variety of investigations.

A National Laboratory in Orbit

In the NASA Authorization Act of 2005 (P.L. 109-155), Congress designated the U.S. segment of the ISS as a National Laboratory, and directed the Agency to seek to increase the utilization of the ISS by other Federal entities and the private sector. Subsequently, in the NASA Authorization Act of 2010 (P.L. 111- 267), Congress directed that the Agency enter into a cooperative agreement with a not-for-profit organization to manage the activities of the ISS National Laboratory. On August 31, 2011, the Agency finalized a cooperative agreement with CASIS to manage the portion of the ISS that operates as a U.S. National Laboratory. CASIS works to ensure that the Station's unique capabilities are available to the broadest possible cross-section of U.S. scientific, technological, and industrial communities. The goal is to support, promote and accelerate innovations and new discoveries in science, engineering, and technology that will improve life on Earth. NASA's National Laboratory partners can use the unique microgravity environment of space and the advanced research facilities aboard Station to enable investigations that may give them the edge in the global competition to develop valuable, high technology products and services. The National Laboratory will help establish and demonstrate the market for research in LEO beyond the requirements of NASA.

Use of the ISS as a National Laboratory has increased significantly since FY 2012, which was the first full year of operations by CASIS. CASIS is reaching its full allocation of National Lab resources and it expected to continue to do so for the foreseeable future. The growth is coming from non-traditional areas, specifically from the commercial sector. Commercial projects for research and technology development on the ISS National Lab have increased from three in FY 2012 to 107 in FY 2014. This includes such industry leaders as Merck, Novartis, and Eli Lilly. Expanded capabilities, such as the ability to conduct model organism research on the ISS, using rodents as well as other organisms, has helped draw this interest. Commercial efforts have also included perhaps the largest purely commercial provision of services using the ISS, the deployment by NanoRacks of dozens of Dove cube satellites for Planet Labs.

Similarly, use by other Government agencies, including the National Institutes of Health and Department of Defense, has also begun to broaden, totaling 11 investigations in FY 2014. Finally, investigations from academic institutions rose from 31 in FY 2012 to 90 in FY 2014. Grant funding for research through the National Lab continues to grow, from \$2.1 million in FY 2012, to \$5.9 million in FY 2014. Additionally, NASA is collaborating with CASIS to enable sustained investment and research activities onboard the ISS across industry and other Government agencies that will transcend the life of the Station. The ISS International Partners are also seeking to expand the base of researchers using their assets on the ISS and are very interested in the National Lab model. This will expand research, and commercial participation, in low-Earth orbit.

ISS – Benefits to Humanity

Almost as soon as the ISS was habitable, researchers began using it to study the impact of microgravity and other space effects. In the physical and biological sciences arena, the ISS allows researchers to use microgravity conditions to understand the effect of the microgravity environment on microbial systems, fluid physics, combustion science, and materials processing, as well as environmental control and fire safety technologies. The ISS also provides a test bed for studying, developing, and testing new technologies for use in future exploration missions. Although each Station partner has distinct agency/national goals for ISS research, each partner collectively shares a unified goal to extend the resulting knowledge for the betterment of humanity. In the areas of human health, telemedicine, education, and Earth observations from space, there are already demonstrated benefits. Pharmaceutical development research, Station-generated images that assist with disaster relief and farming, and education programs that inspire future scientists, engineers, and space explorers highlight just some of the many examples of research that can benefit humanity.

ISS crews are conducting human medical research to develop knowledge in the areas of: clinical medicine, human physiology, cardiovascular research, bone and muscle health, neurovestibular medicine, diagnostic instruments and sensors, advanced ultrasound, exercise and pharmacological countermeasures, food and nutrition, immunology and infection, exercise systems, and human behavior and performance. Many investigations conducted aboard ISS will have direct application to terrestrial medicine. For example, the growing senior population may benefit from experiments in the areas of bone and muscle health, immunology, vestibular response and balance, and from the development of advanced diagnostic systems. The ISS requires telemedicine be used to monitor and treat crews. Optical Computerized Tomography (OCT), fundoscopy, and tonometry are now routinely used onboard the ISS to diagnose and monitor any progression of Visual Impairment Intracranial Pressure (VIIP) syndrome. The ISS Ultrasound aides in the remote diagnosis of a variety of conditions ranging from musculo-skeletal issues and abdominal pains to infection of soft tissues. Similar equipment and techniques can be used on Earth to provide medical care to patients without requiring their travel to a hospital or doctor.

The ISS also plays an important role in promoting education in the science, technology, engineering, and mathematics (STEM) fields, inspiring students to pursue scientific and technical careers. Astronauts aboard ISS participate in educational downlinks with schools, and engage in communicating with people around the world using "ham" radio. The ISS Program also conducts experiments that involve student participation. One example is the Synchronized Position Hold, Engage, Reorient, Experimental Satellites (SPHERES) facility. SPHERES are three bowling-ball sized spherical satellites that are used inside the Station to test telerobotics operations in addition to spacecraft formation flight and autonomous rendezvous and docking maneuvers. NASA, along with the Defense Advanced Research Projects Agency, with implementation by the Massachusetts Institute of Technology, has co-sponsored "Zero Robotics SPHERES Challenge" competitions for high school and middle students from the U.S. and abroad. The competitions challenge students to write software code, which is uploaded to the robots on ISS, and the SPHERES satellites then execute the instructions, in areas such as formation flight and close proximity operations. Student finalists were able to watch their flight program live on NASA-TV.

Conclusion

The ISS has now entered its intensive research and technology demonstration phase. Station will continue to meet NASA's mission objective to prepare for the next steps in human space exploration. Closer to home, NASA's National Laboratory partners can use the unique microgravity environment of space and the advanced research facilities aboard Station to enable investigations that may give them the edge in the global competition to develop valuable, high technology products and services. Furthermore, the demand for access to the ISS enables the establishment of robust U.S. commercial crew and cargo capabilities. Both of these aspects of the U.S. segment of the ISS as a National Laboratory will help establish and demonstrate the market for research in LEO beyond the requirements of NASA.

With NASA as the lead integrator on ISS for the international partnership, the ISS allows the U.S. to demonstrate global leadership in human spaceflight and technology development. ISS and the teams that operate it are an amazing global resource.

NASA appreciates this Committee's ongoing support of the ISS as we work together to support this amazing facility which yields remarkable results and benefits for the world.

Mr. Chairman, I would be happy to respond to any questions you or the other Members of the Subcommittee may have.

In: The International Space Station
Editor: Evelyn Clemens
ISBN: 978-1-63484-071-2

Chapter 4

Statement of Paul K. Martin, Inspector General, NASA. Hearing on "The International Space Station: Addressing Operational Challenges"*

Chairman Babin, Ranking Member Edwards, and Members of the Subcommittee:

The Office of Inspector General (OIG) is committed to providing independent, aggressive, and objective oversight of NASA programs and personnel. Thank you for this opportunity to appear before the Subcommittee today to discuss the challenges NASA faces in operating the International Space Station (ISS or Station), particularly in light of the loss of three cargo resupply flights during the past 8 months. The OIG has issued four reports related to this topic over the past 2 years, including reviews on NASA's plans to extend Station operations until 2024 and its contracts with private companies to fly cargo and eventually crew to the Station.[1] We plan to issue five additional reports related to this topic later this year, including an audit examining the effects of the October 2014 loss of a commercial cargo resupply mission on Station operations.[2]

---

* This is an edited, reformatted and augmented version of a statement presented July 10, 2015 before the House Committee on Science, Space, and Technology, Subcommittee on Space.

The culmination of decades-long international efforts to plan, build, and operate a permanently crewed research platform in space, the ISS is a unique technological achievement that plays a key role in NASA's goal of sending humans to Mars. In November 2013, the ISS completed 15 years of continuous operation in low Earth orbit. With the United States' plan to extend Station operations until 2024, a spacecraft originally designed and tested for a 15-year life span may now operate for 26 years.

While NASA has identified no major obstacles to continued operation of the Station through 2024, in a September 2014 report we found several technical challenges the Agency must address in order to extend ISS operations, including ensuring adequate power generation in light of degradation of the Station's solar arrays.[3] As noted in this report, the number one operational risk for the ISS Program is ensuring the ability to deliver supplies and transport astronauts to and from the Station. The failure of three cargo resupply missions in 8 months – an Orbital Sciences Corporation (Orbital) mission in October 2014, a Russian Progress vehicle in April 2015, and a Space Exploration Technology (Space-X) in June 2015 – underscores the importance of managing this risk. While NASA contracted with two commercial cargo providers to ensure redundancy, both companies' resupply flights are now on hold pending the outcome of accident investigations and approval from the Federal Aviation Administration and NASA to resume cargo missions.

Fortunately, the successful docking of a Russian Progress cargo flight on July 5, 2015, will provide the ISS with much-needed supplies in the short-term until the launch of a Japanese cargo flight in August and NASA's commercial partners return to flight. In the event that Orbital and SpaceX do not return to flight before 2016, NASA may need to make additional operational adjustments, including potentially reducing the number of crew on board the ISS, which in turn would significantly reduce the Station's research capability.

While it appears the ISS has sufficient food and water to sustain the crew for the next several months, NASA lost important supplies in both the October 2014 and June 2015 failed cargo missions. For example, both flights were carrying supplies for the Station's Environmental Control and Life Support System, including replacement parts for the ISS's water purification system. The flights also carried hardware for the first International Docking Adapter that is necessary to support docking operations for the commercial crew vehicles NASA hopes will begin arriving at the Station in 2017. In addition, the Center for the Advancement of Science in Space (CASIS), which manages non-NASA research aboard the ISS, lost more than $675,000 of research on both flights, including 5 stem cell research projects, a microgravity investigation to help fine tune delivery and dosage of drugs, 8 miniature satellites known as Cubesats, and 10 sets of high school and middle school experiments.

Because NASA utilizes the ISS as a research platform to study and mitigate a variety of risks associated with human travel and long-term habitation in space, the Station is an important part of NASA's plans to send humans beyond low Earth orbit. In our judgment, the recent Orbital and SpaceX launch failures have affected research in three ways: (1) a reduction in available crew time due to a temporary delay in returning the Station's crew complement to six, (2) the cost to regenerate some of the research lost, and (3) a delay in the return of experiments due to the suspension of SpaceX flights, which is the only company capable of returning experiments and other cargo to Earth. Moreover, in our September 2014 report on NASA's plans to extend operation of the ISS, we reported that even if Station operations continue through 2024, NASA will not have sufficient time to address all the health and human performance risks for which the Station is an appropriate research platform.[4] Accordingly, NASA needs to prioritize Station research to address the most important risks in the time remaining. Later this year we plan to issue an audit examining in detail NASA's efforts to address the health and human performance risks associated with long-duration space exploration.

In addition to serving as a platform for NASA research, the Station also provides a laboratory for other government agencies and private entities to advance new technologies in health and medicine, robotics, manufacturing, and propulsion. In August 2011, NASA signed a cooperative agreement with CASIS to manage non-NASA research aboard the ISS. Pursuant to the agreement, NASA provides CASIS $15 million annually and the organization is expected to raise additional funds from private entities and encourage companies to self-fund research. When we interviewed CASIS officials as part of our ISS extension audit, they told us that provisions in the agreement with NASA requiring researchers to assign patent licenses and data rights to the Government were deterring commercial stakeholders from conducting research on the ISS. To address this issue, NASA submitted proposed legislation to Congress in June 2013 that would allow researchers to retain "all rights in inventions made... during the conduct of [Station] activities." To date, the legislation has not moved forward.

While utilization of the ISS for research has increased, several factors continue to pose limits to fully utilizing the Station. First, until a seventh crew member is brought onboard, NASA will not be in a position to maximize the amount of crew time dedicated to research on the Station.[5] Moreover, during 2016 substantial crew time will be devoted to reconfiguring the ISS to accommodate the commercial vehicles NASA hopes will be ready to transport astronauts in 2017.

To that point, securing safe and reliable crew transportation remains a major operational challenge to ensuring the continued effectiveness of the ISS. The fourth and final phase of NASA's Commercial Crew Program began in September 2014 with the award of $6.8 billion in firm-fixed-price contracts to The Boeing Company ($4.2 billion) and SpaceX ($2.6 billion) to complete development and certification of their respective spaceflight systems. Under these contracts, NASA will provide specific requirements for launch systems, spacecraft, and related ground support. The contracts include at least one crewed flight test with a NASA astronaut to verify that the fully integrated rocket and spacecraft system can launch, maneuver in orbit, and dock to the ISS, as well as to validate that all systems are performing as expected. Upon certification, each company will conduct at least two, but as many as six, crewed missions to the Station.

In a November 2013 audit report, we identified four challenges to NASA's Commercial Crew Program: (1) unstable funding, (2) integration of cost estimates with the Program schedule, (3) providing timely requirement and certification guidance to contractors, and (4) spaceflight coordination issues with other Federal agencies.[6] Since that time, NASA has made progress in these areas and expects to address our report recommendations by the end of this month. In May 2015, we began a follow-on audit examining whether the Commercial Crew Program is meeting its planned cost and schedule goals and how it is managing risks and certification requirements.

NASA's annual cost to operate the ISS in fiscal year 2014 was almost $3 billion. Those costs included on-orbit vehicle operations, research, crew transportation, and cargo resupply by U.S. commercial and international partner vehicles. During fiscal year 2016, the ISS Program plans to spend $1.1 billion (almost 36 percent of its budget) on operation and maintenance of the Station and another $1.1 billion on cargo transportation.

Going forward, NASA officials have indicated they intend to maintain the ISS budget between $3 and $4 billion per year through 2024; however, we suspect the cost to NASA will likely be higher. First, much of the projected cost increase can be attributable to higher transportation costs, but we found NASA's estimates for cargo and crew transportation to be optimistic. Second, the Agency's international partners have yet to commit to participating in Station operations beyond 2020. A decision from one or more not to participate could drive up costs for NASA. Given the importance of international cooperation to the ISS and other NASA missions, the OIG is conducting an audit examining NASA's efforts to partner with the space agencies of other countries on science and exploration-related projects.

In addition, as a follow-on to our work on extending the Station to 2024, we are examining NASA's ISS contracts to assess whether the Agency is doing everything possible to avoid incurring unnecessary costs. In the course of this audit, we learned that NASA has taken actions to reduce the operations and maintenance costs of the ISS Program, including competing contracts and taking initial steps to convert portions of the largest ISS contract to a fixed-price vehicle. However, at this point it is unclear to what extent these strategies will result in future cost savings.

In conclusion, the OIG looks forward to continuing our cooperative relationship with NASA, this Subcommittee, and other congressional committees as we examine NASA's effort to address challenges in effectively operating the ISS.

End Notes

[1] NASA OIG, "Extending the Operational Life of the International Space Station Until 2024," (IG-14-031, September 18, 2014); "NASA's Management of the Commercial Crew Program," (IG-14-001, November 13, 2013); "NASA's Efforts to Maximize Research on the International Space Station (ISS)," (IG-13-019, July 8, 2013); and "Commercial Cargo: NASA's Management of Commercial Orbital Transportation Services and ISS Commercial Resupply Contracts," (IG-13-016, June 13, 2013).

[2] In addition to a report on the October 2014 mission failure, we are examining NASA's management of ISS contracts, NASA's efforts to manage health and human performance risks for space exploration, challenges to international cooperation in space programs, and NASA's Commercial Crew Program. We also plan to open a review examining the impact of the loss of the June 2015 commercial resupply mission.

[3] NASA OIG, Extending the Operational Life of the ISS.

[4] NASA OIG, Extending the Operational Life of the ISS.

[5] Although the ISS can support a crew of seven, currently only six individuals can be on Station at one time to accommodate evacuation in case of an emergency. The Russian capsule, which is currently the only vehicle transporting astronauts to the Station, has a three-person capacity and only two Soyuz capsules can be attached to the Station simultaneously.

[6] NASA OIG, Commercial Crew Program.

In: The International Space Station
Editor: Evelyn Clemens

ISBN: 978-1-63484-071-2

Chapter 5

INTERNATIONAL SPACE STATION: CHALLENGES TO INCREASED UTILIZATION MAY AFFECT RETURN ON INVESTMENT. STATEMENT OF SHELBY S. OAKLEY ACTING DIRECTOR, ACQUISITION AND SOURCING MANAGEMENT, GOVERNMENT ACCOUNTABILITY OFFICE. HEARING ON "THE INTERNATIONAL SPACE STATION: ADDRESSING OPERATIONAL CHALLENGES"*

United States Government Accountability Office

* This is an edited, reformatted and augmented version of a report, GAO-15-722T, issued by the U.S. Government Accountability Office. The report was originally presented as testimony July 10, 2015 before the House Committee on Science, Space, and Technology, Subcommittee on Space.

Why GAO Did This Study

The United States has spent tens of billions of dollars to develop, assemble, and operate the ISS over the past two decades. NASA plans to spend about $22 billion more from fiscal year 2016 through 2020—with over half of that planned for transportation—on the ISS. In January 2014, the Administration proposed extending the life of the ISS to at least 2024 to take further advantage of the investment in the ISS. Since 2005, Congress enacted several laws to increase utilization of the ISS by commercial and academic researchers. The NASA Authorization Act of 2010 required NASA to enter into a cooperative agreement with a not-for-profit entity to manage the ISS National Laboratory and in 2011 it did so with CASIS. CASIS is charged with maximizing use of the ISS for scientific research by executing several required activities.

This statement will provide an overview of (1) NASA's budget for ISS and the factors affecting budget levels through 2020, (2) several challenges that could impact effective utilization of ISS by both NASA and CASIS, and (3) steps that NASA and CASIS could take to better document and assess CASIS's progress in this regard.

This statement is based primarily on GAO's April 2015 report (GAO-15-397) as well as other prior reports and testimonies. GAO also conducted a limited amount of additional audit work in June 2015 to update certain information.

What GAO Found

Based on GAO analysis of the National Aeronautics and Space Administration's (NASA) fiscal year 2016 budget estimate, the agency anticipates that the costs to operate, sustain, perform research, and provide crew and cargo transportation to the International Space Station (ISS) are projected to increase by almost $1 billion—or almost 53 percent—from fiscal year 2015 to fiscal year 2020 when the projected costs are expected to exceed $4 billion. The majority of the total projected cost increase for ISS is attributable to commercial crew and cargo transportation. The budget for ISS cargo and crew transportation is currently planned to increase by over $700 million from fiscal year 2016 to fiscal year 2020—or over 55 percent of the total ISS budget—which includes the purchase of six Russian Soyuz seats in

2018 and commercial crew missions beginning in fiscal year 2019. The costs to operate the ISS and perform research are expected to be stable with only slight increases through fiscal year 2020.

NASA and the Center for the Advancement of Science in Space (CASIS)—a non-profit entity selected to managc non-NASA research on the ISS National Laboratory—must overcome several challenges to increase utilization and achieve a better return on investment. NASA and CASIS officials told GAO that the ISS will be challenged to meet an expected increase in demand for crew time and certain research facilities. Securing cargo transportation has also presented challenges. CASIS-sponsored researchers have experienced cost increases of almost $500,000 because of a cargo resupply launch failure in October 2014 and delays to other cargo resupply missions. GAO found that absorbing the increased cost has been a challenge for CASIS given its limited research budget and it could be faced with additional cost increases given the June 2015 launch failure of another cargo resupply mission.

In April 2015, GAO found that CASIS had taken steps to manage and promote research activities on the ISS National Laboratory, but that CASIS and NASA could do more to objectively define, assess, and report progress toward increased utilization. While CASIS had established annual metrics, it did not establish measurable targets for these metrics. GAO has previously reported that performance metrics should have quantifiable targets to help assess whether overall goals are achieved. Consequently, GAO recommended that the ISS program and CASIS develop measurable targets for CASIS's metrics for fiscal year 2016 and beyond. NASA concurred with this recommendation and indicated that these targets should be established by the end of 2015. GAO's April 2015 report also found that while NASA performs an annual assessment of CASIS's performance, the assessment is not documented. This type of documented information can support future assessments of return on investment. GAO recommended that NASA document the annual program assessment of CASIS performance. NASA concurred with this recommendation and plans to take action in response to CASIS's 2015 annual report. Because CASIS is allocated at least 50 percent of ISS research capacity, ensuring that CASIS continues to make progress promoting research activities and achieving its goal to increase utilization of ISS is essential to demonstrate a return on investment for the tens of billions of dollars already invested and that will continued to be invested in ISS.

Chairman Babin, Ranking Member Edwards, and Members of the Subcommittee:

Thank you for the opportunity to discuss our work on the National Aeronautics and Space Administration's (NASA) management of the International Space Station (ISS). The United States has spent tens of billions of dollars over the past two decades to develop, assemble, and operate the ISS, which has been used as a manned research outpost continuously for over 14 years. NASA plans to spend about $22 billion more from fiscal year 2016 through 2020—with over half of that planned for transportation—to enable further scientific research the agency views as critical to future human space activities. In January 2014, the Administration proposed extending the life of the ISS by a minimum of 4 years to at least 2024 to take further advantage of the investment in the ISS. Congress enacted several laws to increase utilization of the ISS by commercial and academic researchers.[1] In response to direction in the NASA Authorization Act of 2010, in 2011 NASA selected the Center for the Advancement of Science in Space (CASIS), a non-profit entity, to manage non-NASA commercial and academic research aboard the ISS National Laboratory. Because CASIS is allocated at least 50 percent of ISS research capacity and was created to maximize the value of the ISS investment, future success of the ISS as a research platform is partially dependent on CASIS's success.

My statement today will provide an overview of NASA's budget for ISS and the factors affecting budget levels through 2020. In addition, my statement will also focus on several challenges that could impact effective utilization of the ISS by both NASA and CASIS as they continue their efforts to demonstrate that the research and technology development performed aboard the ISS National Laboratory benefits life on Earth and develop commercial markets that can be sustained in low-Earth orbit. Finally, I will discuss steps that NASA and CASIS could take to better document and assess CASIS's progress in this regard.

In preparing this statement, we primarily relied on work conducted to support our April 2015 report.[2] Additionally, we relied on our prior reports and testimonies, including those related to CASIS's management of the ISS National Laboratory, the agency's acquisition approach for commercial crew transportation, and ISS sustainment and utilization.[3] Information on our scope and methodology is available in the reports cited in this statement. We also conducted a limited amount of additional audit work in June 2015 to update information on CASIS's efforts to increase utilization of the ISS and planned commercial crew and cargo partner missions. The work upon which this statement is based was performed in accordance with generally accepted government auditing standards. Those standards require that we plan and perform the audit to obtain sufficient and appropriate evidence to provide a reasonable basis for our findings and conclusions based on our audit objectives. We believe that the evidence obtained provides a reasonable basis for our findings and conclusions based on our audit objectives.

BACKGROUND

The ISS supports research projects with state of the art facilities for Earth and space science, biology, human physiology, physical science, and materials research, and provides a platform to demonstrate new space-related technologies. The ISS currently has three crew members in the U.S. operating segment who, according to NASA officials, devote a total of approximately 35 hours per week to conduct research. The remaining crew time is used for operations and maintenance of the ISS, training, exercise, and sleep.

Congress Directs Changes in Management of the ISS

Since 2005, Congress has directed several changes regarding the management and utilization of the ISS. For example, the NASA Authorization Act of 2005 designated the U.S Operating Segment of the ISS as a national laboratory and the NASA Authorization Act of 2010 directed the NASA Administrator to provide initial financial assistance and enter into a cooperative agreement with a not-for-profit organization to manage the activities of the ISS National Laboratory for non-NASA utilization of the ISS research capabilities and available facilities.[4] The 2010 act also requires the ISS National Laboratory-managed experiments to be guaranteed access to and use of at least 50 percent of the U.S. research capacity allocation including power, facilities to keep experiments cold, and requisite crew time onboard the ISS through September 30, 2020. Our April 2015 report provides a synopsis of these legislative actions.[5]

In August 2011, after a competitive process, NASA signed a cooperative agreement with CASIS, a not-for-profit entity, to manage the activities of the ISS National Laboratory through September 30, 2020.[6] CASIS is bound by the responsibilities outlined in the cooperative agreement, which tasks it with maximizing the value of the ISS National Laboratory by stimulating interest and use of the ISS for scientific research by directly soliciting potential users and fostering a market to attract others. CASIS is also charged with maximizing the use of the ISS for advancing science, technology, engineering, and mathematics education. Pursuant to the cooperative agreement, NASA will provide CASIS $15 million annually through 2020, of which it will seek to award at least $3 million in research grants. CASIS officials have stated that the remainder of NASA funding is used for infrastructure and direct costs such as labor and travel-related expenses. Further, CASIS is also responsible for soliciting non-NASA funding for research by targeting various sources.

NASA Is Taking Steps to Establish Commercial Transportation Capability

In 2005, NASA established the Commercial Crew and Cargo Program to facilitate the private demonstration of safe, reliable, and cost-effective transportation services to low-Earth orbit and encourage innovation in the private sector. The goal of this program is to enable the government to buy both cargo and crew commercial transportation services at a reasonable price. NASA is procuring cargo transportation to the ISS through a commercial resupply services contract that was signed with Orbital Sciences Corporation (Orbital) and Space Exploration Technologies Corporation (SpaceX) in 2008. As of April 2015, SpaceX has launched six successful resupply missions and Orbital has launched two successful resupply missions. Orbital and SpaceX are scheduled to provide 8 and 15 resupply flights, respectively, through December 2017, although the number of Orbital flights may be modified. Orbital resupply flights to the ISS were deferred pending a review of a launch failure that occurred during a resupply launch in October 2014, which resulted in the loss of that mission. According to NASA officials, both Orbital and NASA are still reviewing the mishap and have not provided final reports as of June 2015. The ISS program anticipates that Orbital will resume missions by the end of 2015; however, the flight will use an Atlas rocket to launch a Cygnus spacecraft instead of Orbital's Antares launch vehicle. In late June 2015, SpaceX experienced a launch failure during a cargo resupply launch that resulted in the loss of that mission. NASA is currently evaluating the impact this recent launch failure will have on the ISS program. In September 2014, NASA released a request for proposals for its Commercial Resupply Services 2 (CRS2) requirements, which would result in a follow on contract to the

Commercial Resupply Services contracts that were awarded to SpaceX and Orbital in December 2008. NASA expects to award CRS2 contracts in September 2015 for cargo transportation services beginning in 2018, about 4 months later than anticipated.

NASA has relied upon Russia to provide crew transportation to and from the ISS since the retirement of the Space Shuttle in 2011. NASA has purchased crew launches from Russia through 2017 and crew rescue and return through mid-2018. NASA purchases seats—at a cost of over $65 million each in 2015—aboard Russian Soyuz space capsules. To support its goal of obtaining low cost domestic crew transport, in September 2014, NASA awarded contracts to The Boeing Company (Boeing) and SpaceX to develop a capability to transport astronauts to and from the ISS. Those awards include a minimum of two to a maximum of six crewed service missions per provider. The Commercial Crew Program, which is outside of the ISS program, is currently responsible for the cost to develop this capability and will also fund one crewed service flight per provider. ISS program officials have stated that they expect the first service flights to take place in fiscal year 2018. The ISS program will fund the remaining service flights. NASA expects this capability to be available in fiscal year 2018; however, it is unclear whether those vehicles currently in development under the commercial crew program will be ready in time. Once this capability has been established, NASA plans to increase the number of astronauts in the U.S. operating segment of the ISS from three to four.

ISS Budget Projected to Rise to Account for Increasing Transportation Costs while Operations and Research Costs Remain Stable

NASA makes a significant investment in the ISS program each year. Based on our analysis of NASA's fiscal year 2016 budget estimate, the agency anticipates that the costs to operate, sustain, perform research, and provide crew and cargo transportation to the ISS are projected to increase by almost $1 billion—or almost 53 percent—from fiscal year 2015 to fiscal year 2020 when the projected costs are expected to exceed $4 billion. See figure 1 for ISS funding from fiscal year 2010 through fiscal year 2020.

The majority of the total projected cost increase for ISS is due to the ISS program's need to pay for commercial crew and cargo transportation. The budget for ISS cargo and crew transportation is currently planned to increase by over $700 million from fiscal year 2016 to fiscal year 2020, at which point it will comprise over 55 percent of the total ISS budget.

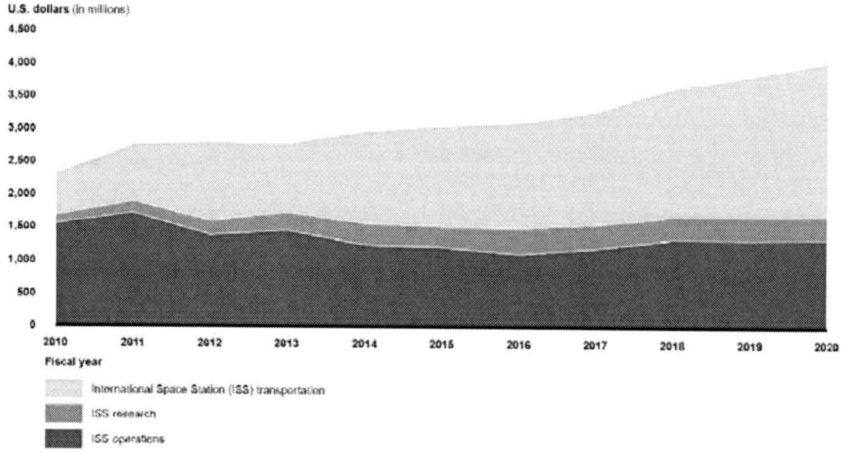

Source: GAO presentation of NASA data. | GAO-15-722T.

Figure 1. International Space Station Funding by Budget Element for Fiscal Years 2010 through 2020.

ISS program officials told us that in fiscal year 2017, the program will begin to fund commercial crew missions that are expected to take place in fiscal year 2019.[7] NASA has also initiated steps to purchase six Soyuz seats from Russia for flights to the ISS in 2018, the cost of which ISS program officials said was accounted for in the projected transportation costs. If NASA determines that domestic commercial entities are able to fulfill crew transportation requirements in 2018, those vehicles will become NASA's primary transportation source to the ISS and the Soyuz seats purchased may then be utilized as backup transportation or to augment future ISS needs.

Based on our analysis of NASA's fiscal year 2016 budget estimate, the cost to operate the ISS is expected to be relatively stable with only slight increases through fiscal year 2020. The ISS operations costs decreased $500 million—or 30 percent—from a peak in fiscal year 2011 through fiscal year 2015. Operations costs are expected, however, to increase by approximately $130 million from fiscal year 2017 through fiscal year 2020, which NASA officials attribute in part to inflation and the addition of the fourth crew member. Our past work on Department of Defense (DOD) aircraft systems similarly found that operations costs can increase over time.[8] There may be other factors that could increase operations costs for the ISS over time such as the need for additional spare parts and mitigations needed for structural issues. The ISS program has implemented a number of initiatives that have yielded cost savings or containment. For example, NASA reduced operations costs by scaling back ISS program and contractor workforce levels and by combining several contracts. The NASA Inspector General currently has ongoing work assessing NASA's efforts to combine and consolidate ISS contracts for operations and maintenance.

NASA also projects that the cost for the ISS Research account will remain stable through fiscal year 2020, when research costs are expected to be about 9 percent of total ISS projected costs. Within this account, NASA provides limited funding to CASIS—$15 million per year—of which a minimum of $3 million is to be used to sponsor non-NASA research aboard the ISS by commercial, academic, and other government agency users. Although NASA research costs are projected to remain stable, CASIS sponsored research is not limited to this $3 million minimum. For example, our analysis of CASIS information shows that CASIS has averaged more than $4.3 million in grants paid out to researchers each fiscal year since 2012 and has paid over $15 million to its grantees through March 2015.

NASA and CASIS Face Challenges to Increase Utilization and Sustain the ISS Which Could Affect Return on Investment

As we reported in April 2015, NASA and CASIS must overcome several challenges to increase utilization and sustain the ISS until 2024 and achieve a better return on the investment. According to NASA and CASIS officials, as CASIS increases the number of experiments for the ISS National Laboratory, the demand for crew time and certain research facilities aboard the ISS is expected to increase and officials project the ISS National Laboratory will be challenged to meet that demand. NASA officials told us that while the demand for crew time is currently manageable, it remains allocated at or near 100 percent, as the three crew members on the U.S. segment of the ISS utilize most of the 35 hours scheduled per week to conduct research. Crew time for research is expected to double on the ISS National Laboratory once the crew increases from three to four astronauts in fiscal year 2018 because, according to NASA officials, the additional crew member will devote most of his or her time to research. NASA's ability to support increased utilization through an additional crew member, however, is reliant on commercial crew providers providing the promised capability and NASA's ability to fund the effort. Both Boeing and SpaceX plan to hold demonstration flights to the ISS in 2017, but risks remain for both contractors.

According to CASIS officials, they have been challenged to raise additional funding from external sources to supplement the amount of funding provided by NASA. CASIS officials attributed this challenge to the fact that CASIS is a new non-profit entity and the value of performing research aboard the ISS has not been fully demonstrated. Although CASIS's business development team is actively identifying partnerships and funding opportunities with commercial and non-profit granting organizations, CASIS officials said that it takes time to identify, develop, and mature these partnerships to result in funding support. Through December 2014, CASIS reported that it had received funding commitments from external sources of approximately $12 million to support its research mission. However, according to CASIS's fiscal year 2014 annual report, published in December 2014, CASIS received contributions totaling only $9,193 in 2014. NASA officials stated that doing research aboard the ISS National Laboratory can take upwards of 2 to 3 years to plan and execute, timelines that are generally not acceptable to commercial companies that desire a more rapid return on their investments. CASIS and NASA officials also told us that the value of doing research aboard the ISS National Laboratory has to be further demonstrated so commercial industries can be convinced it is worth the high investment.

We also reported in April 2015 that the ability to secure cargo transportation for selected research investigations to the ISS is outside of CASIS's control and has presented challenges. NASA provides launch services to the ISS National Laboratory through its commercial resupply services contracts and CASIS receives cargo allocations for its sponsored research. Budget shortfalls due in part to the effects of sequestration and the Orbital launch failure have resulted NASA cancelling 4 of 46 planned cargo flights through 2020. In January 2015, NASA's Aerospace Safety Advisory Panel noted that both commercial cargo launch providers had struggled to meet desired launch dates and that the schedule performance must significantly improve to enable consistent scientific research on the ISS.[9] The panel added that there will be additional pressure on cargo logistics while Orbital works through its plan to resume cargo missions. This pressure will likely be increased because of the June 2015 launch failure of a SpaceX cargo resupply mission. We found that such launch failures and delays have resulted in cost increases for CASIS-sponsored researchers. For example, the rocket launch failure to the ISS in October 2014 resulted in the loss of several CASIS-sponsored research investigations at a total cost of almost $175,000 which includes hardware and materials, labor consulting, and grants. In addition, launch delays for another cargo resupply mission resulted in over $300,000 in cost increases for several researchers. This included costs for additional materials and samples such as biological payloads that have a limited viability or very specific requirements associated with the timing of the payload flight and often require consumables such as gas and water that must be replenished when a launch is delayed. Absorbing the increased cost has been a challenge for CASIS given its limited research budget, but it is addressing the cost increases due to delays by asking researchers that have biological payloads to identify the impact and associated costs for launch delays in their budgets so it can plan for budget reserves, as necessary.

To extend the ISS service life to 2024 or beyond, NASA must ensure that spare parts are sufficient and available when needed and that the structures are sound. In December 2011, we found that NASA had a reasonable approach to determine, obtain, and deliver necessary spare parts to the ISS through 2020.[10] At that time, we found that NASA had given equal weight to manufacturers' predictions and actual performance. However, because NASA has generally found failure rates for replacement units to be lower than manufacturers' predictions, over time the resulting estimates could prove to be overly conservative. More recently, the ISS program reported that one part was failing at rates greater than projected and that spares were not available as a result. In December 2011, we recommended that NASA should reassess the relative weight given to original reliability estimates of spares' life expectancies as performance data accumulates. NASA concurred with this recommendation and the ISS program has taken action to revisit to the methods it uses to calculate the need for spare parts for the ISS. NASA performs an annual assessment of the spare parts, which has resulted in the procurement of spare parts based in part on actual hardware performance. In our December 2011 report, we also found that NASA is using reasonable analytical tools to assess structural health and determine whether ISS hardware can operate safely through 2020. On the basis of prior analysis of structural life usage through 2015 and the robust design of the ISS structures, NASA anticipated that—with some mitigation—the ISS will remain structurally sound for continued operations through 2020. At the time of our 2011 review, NASA had assessed only 40 percent, by weight, of the assembled ISS because most of the ISS structures have not been on orbit long enough to accumulate the data needed for analysis. NASA expected to complete ISS structural assessments in early 2016. These are positive steps; however, continued efforts such as this will be important to ensure that processes NASA uses to evaluate the need for spare parts are adequate to operate and sustain the ISS for at least an additional 4 years.

OBJECTIVE ASSESSMENT OF PROGRESS NEEDED TO DEMONSTRATE RETURN ON INVESTMENT

Despite these challenges, in April 2015, we reported that CASIS had taken steps to carry out its responsibilities to manage and promote research activities on the ISS National Laboratory as outlined in its cooperative agreement.[11] For example, CASIS had identified key research areas and released seven requests for proposals to solicit interest for research projects. While we noted this progress, we found that CASIS and NASA could do more to objectively define, assess, and report progress toward increased utilization. Specifically, we found that while CASIS had established annual metrics that met most of the key attributes of successful performance measures, it did not establish measurable targets or goals for either fiscal year 2014 or 2015 metrics. We have previously reported that performance metrics should have quantifiable, numerical targets or other measurable values, which help assess whether overall goals and objectives were achieved.[12] We concluded that without these targets, NASA and CASIS cannot conduct assessments of CASIS's efforts to increase ISS utilization that are objective, measurable, or conclusive. To enable such assessments, in April 2015 we recommended that the ISS program manager work with CASIS to collectively develop and approve measurable targets for CASIS's metrics for fiscal year 2016 and beyond. NASA concurred with this recommendation and indicated that these targets should be established by December 31, 2015.

Also in our April 2015 report, we found that while NASA performs an annual assessment of CASIS's performance, the assessment is not documented. Federal standards for internal control call for information to be recorded and communicated to management and others who need it to carry out their responsibilities. This type of documented information is important to support decision making and to support future assessments of return on investment. However, without definitive and documented assessment factors,

NASA will also be challenged to take action in response to CASIS performance. According to the cooperative agreement between NASA and CASIS, continued funding of CASIS is contingent on the scientific progress of the project and that NASA will assess such progress in fiscal year 2020 to make a determination about whether to extend or terminate the cooperative agreement.[13] We concluded that without documentation based on objective measures of performance, NASA lacks support to make such a decision. We therefore recommended that the ISS program manager document the annual program assessment of CASIS performance in order to provide CASIS management actionable information to better fulfill its responsibilities. NASA concurred with the recommendation and stated that officials would begin documenting the agency's annual program assessment in response to CASIS's 2015 annual report. It will be important that NASA and CASIS follow through on their commitments to have objective measures that will enable NASA to measure CASIS's progress to increase utilization and demonstrate the return on the investment of the ISS.

In conclusion, the ISS offers the potential for scientific breakthroughs, a unique test bed for new technologies and applications, and a platform for increased commercial and academic research and NASA has made an important commitment to the future of research aboard ISS by proposing to extend operations to 2024. Key to supporting this commitment is effectively managing the challenges that could affect efforts to maximize the return on investment for the tens of billions of dollars that have been spent on the ISS. Achieving greater utilization of the ISS and its unique capabilities, showing the benefit of commercial and academic research, and demonstrating success to generate increased interest from potential users could help NASA demonstrate such a return. Because CASIS is allocated at least 50 percent of ISS research capacity, ensuring that CASIS continues to make progress promoting research activities and achieving its goal to increase utilization of the ISS is essential. Even with an extension of operations to 2024, CASIS has limited time to demonstrate that the research and technology development performed aboard the ISS National Laboratory benefits Earth and commercial markets can be sustained in low-Earth orbit. By NASA and CASIS working together in the coming years to address challenges that could negatively affect increased utilization of this unique research facility and to identify and document objective measures of success to demonstrate a return on investment, such a return might be realized.

Chairman Babin, Ranking Member Edwards, and Members of the Subcommittee, this completes my prepared statement. I would be pleased to respond to any questions that you may have at this time.

End Notes

[1] Commercial Space Act of 1998, Pub. L. No.105-303, § 101; National Aeronautics and Space Administration Authorization Act of 2005, Pub. L. No.109-155, § 507; and National Aeronautics and Space Administration Authorization Act of 2010, Pub. L. No. 111-267, § 504.

[2] GAO, International Space Station: Measurable Performance Targets and Documentation Needed to Better Assess Management of National Laboratory, GAO-15-397 (Washington, D.C.: April 27, 2015).

[3] GAO, NASA: Significant Challenges Remain for Access, Use, and Sustainment of the International Space Station, GAO-12-587T (Washington, D.C.; March 28, 2012); International Space Station: Approach for Ensuring Utilization through 2020 Are Reasonable but Should be Revisited as NASA Gains More Knowledge of On-Orbit Performance, GAO-12-162 (Washington, D.C.; December 15, 2011); and National Aeronautics and Space Administration: Acquisition Approach for Commercial Crew Transportation includes Good Practices, but Faces Significant Challenges, GAO-12-282 (Washington, D.C.: December 15, 2011).

[4] Pub. L. No.109-155, § 507 and Pub. L. No. 111-267, § 504.

[5] GAO-15-397.

[6] The Administration recently proposed extending the operational life of the ISS from 2020 to at least 2024. Our prior work has shown that it is technically feasible to extend the ISS operational life to at least 2028.

[7] According to ISS officials, the Commercial Spaceflight Program will fund the first flight for Boeing and SpaceX and the ISS program will then fund the second and any subsequent crewed flights.

[8] GAO, Defense Management: DOD Needs Better Information and Guidance to More Effectively Manage and Reduce Operating and Support Costs of Major Weapons Systems, GAO-10-717 (Washington, D.C.: July 20, 2010).

[9] National Aeronautics and Space Administration, NASA Aerospace Safety Advisory Panel Annual Report for 2014 (Washington, D.C., Jan. 28, 2015). The Aerospace Safety Advisory Panel was established by Congress in 1968 to provide advice and make recommendations to the NASA Administrator on safety matters. The panel reviews safety studies and operations plans and advises the NASA Administrator and Congress on hazards related to proposed or existing facilities and operations, safety standards and reporting, safety and mission assurance aspects regarding ongoing or proposed programs, and NASA management and culture issues related to safety.

[10] GAO-12-162.

[11] GAO-15-397.

[12] GAO, Tax Administration: IRS Needs to Further Refine Its Tax Filing Season Performance Measures, GAO-03-143 (Washington, D.C.: Nov. 22, 2002).

[13] The current cooperative agreement includes a provision for extension beyond 2020.

In: The International Space Station
Editor: Evelyn Clemens
ISBN: 978-1-63484-071-2

Chapter 6

STATEMENT OF JAMES A. PAWELCZYK, ASSOCIATE PROFESSOR OF PHYSIOLOGY, KINESIOLOGY AND MEDICINE, PENNSYLVANIA STATE UNIVERSITY. HEARING ON "THE INTERNATIONAL SPACE STATION: ADDRESSING OPERATIONAL CHALLENGES"*

ABSTRACT

During this decade of International Space Station (ISS) operations, NASA has made enormous strides to develop and implement a research program that will take humans to Mars. The evolving exploration architecture incorporates a space life sciences strategy aligned with the National Research Council's recent Life and Physical Sciences Decadal study. Research remains constrained by competing priorities, limited funding, available crew time, and powered up- and down-mass. To capitalize on the remaining life of the ISS, and to keep the United States at the forefront of exploration, a robust ground-based research program that fully engages the help of the external science community must be aligned with a flight research program designed to keep humans healthy in fractional gravity environments for periods of time exceeding a year. By doing so we can achieve the penultimate goal of the ISS program; to

* This is an edited, reformatted and augmented version of a statement presented July 10, 2015 before the House Committee on Science, Space, and Technology, Subcommittee on Space.

endow future space explorers with the knowledge, skills and abilities to operate independently from Earth.

Mr. Chairman and Members of the Sub-Committee:

Good morning. I thank you for the opportunity to discuss the status of research using the International Space Station. I have been a space life sciences researcher for more than 25 years, regularly funded by grants from NASA. From 1996-1998 I took leave from my academic position at The Pennsylvania State University to serve as a payload specialist astronaut, or guest researcher, on the STS-90 Neurolab Spacelab mission, which flew on the space shuttle Columbia in 1998. I have more than 15 years of experience advising NASA on its life sciences strategy and portfolio, either as a direct consultant or through committees of the National Academies of Science, Engineering and Medicine. I help evaluate NASA's Bioastronautics Research Program for the Institute of Medicine. I am also inaugural member of the National Research Council's (NRC) newly constituted Committee on Biological and Physical Sciences in Space (CBPSS). Part of our charge is to monitor NASA's progress in implementing the recommendations contained in, "Recapturing a Future for Space Exploration: Life and Physical Sciences Research for a New Era," published by the NRC in 2011[1].

The ISS provides a unique platform for research. Past NRC studies have noted the critical importance of the ISS's capabilities to support the goal of long-term human exploration in space. These capabilities include the ability to perform experiments of extended duration, the ability to continually revise experiment parameters on the basis of previous results, the flexibility in experimental design provided by human operators, and the availability of sophisticated experimental facilities with significant power and data resources. The ISS is the only platform of its kind, and it is essential that its presence and dedication to research for the life and physical sciences be fully employed for as long as it is practicable to do so.

To prepare for this hearing, you asked four specific questions:

1. What are the opportunities and challenges in conducting space life and physical science research on the ISS and what should be done to address them?
2. What are some of the most critical areas of ISS research in space life and physical sciences to enabling the long-term goal of sending

humans to the surface of Mars, and what is the status of progress on that research?

3. How are priorities for research on the ISS established and is there a clear and well understood process for aligning ISS resources with those priorities?
4. What are the implications of the proposed extension of ISS operations to 2024 on research and what criteria should Congress use to consider the proposed extension?

In the time allotted, I'd like to share my generally positive view of NASA's progress, and provide some specific suggestions to maximize the use of this extraordinary national resource that has been orbiting our planet every 90 minutes for the past 17 years. My comments will not stray far from my areas of expertise in the life sciences, but many of them should be applicable to the physical sciences as well.

1. What Are the Opportunities and Challenges in Conducting Space Life and Physical Science Research on the ISS and What Should Be Done to Address Them?

The 2009 report from the Review of U.S. Human Spaceflight Plans Committee (the "Augustine Commission") emphasized that future astronauts will face three unique stressors[2]:

- prolonged exposure to solar and galactic radiation;
- prolonged periods of exposure to microgravity; and,
- confinement in close, relatively austere quarters along with a small number of other crew members who must live and work as a cohesive team for many months while having limited contact with their family, friends and culture.

All of these stressors are present in the ISS environment. Martian operations add more stressors: a dusty, dim, energetic environment and a gravitational field that is a little more than a third of our own. Research to address the biological response to fractional gravity is perhaps the area most impacted by changes to the ISS program over the decades. Unless we improve our research centrifuge capabilities on the ISS, we accept a risk of sending humans to Mars with little or no knowledge of how mammalian biology responds in a gravitational field other than Earth's.

My colleagues in the science community report that two of the major challenges to the biology research portfolio are limited access to the ISS and limited crew time. Some types of research, particularly that employing small mammals, is very time consuming to execute. Animal husbandry for a single rodent experiment can easily outstrip available ISS crew time for research during an increment. We can reasonably anticipate that competition for crew time will become worse as the facility ages, and demands on crew time to perform necessary maintenance become more acute.

Access to the ISS for research is not just a matter of access to space, it is a matter of competing programs. ISS research time is allocated in roughly equal proportions between NASA sponsored, peer-reviewed science and projects sponsored by the Center for the Advancement of Science in Space (CASIS), regardless of what that research might be. The outcome is that National Laboratory research and peer-reviewed, NASA-sponsored research vie for scarce resources such as crew time and positions on the flight manifest; in some cases forcing NASA research to lower-fidelity Earth-based analogs such as bed rest research for muscle atrophy and bone demineralization.

The extension criteria report requested by Congress in the NASA Authorization Act of 2015 creates opportunities to better coordinate NASA and CASIS sponsored research. For example, the ISS Program Office could require an experimental definition phase to maximize science return by combining compatible experiments and expanding biospecimen-sharing experiments to answer the most pressing research questions.

2. What Are Some of the Most Critical Areas of ISS Research in Space Life and Physical Sciences to Enabling the Long-Term Goal of Sending Humans to the Surface of Mars, and What Is the Status of Progress on That Research?

The biological risks associated with exploration-class spaceflight are far from being mitigated. This conclusion is based on analysis of 40 years of NASA-sponsored research.

Since the days of Skylab, NASA-funded investigators conducted an aggressive and successful biological research program that was robust, comprehensive, and internationally recognized. Beginning with those early efforts, and continuing with our international partners on the *Mir* and the ISS, we have built a knowledge base that defines the rate at which humans adapt during spaceflight up to six-months duration, with four data points exceeding one-year duration. Right now, we are expanding the one-year database! To prepare for Mars, we need to extend the duration further – up to three years - using a combination of astronaut volunteers and small mammals such as rats and mice.

In *Life of Reason*[3], George Santayana warned that, "those who cannot remember the past are condemned to repeat it." We should not forget the precipitous drop in NASA-sponsored research in the first decade of the millennium. The 2001 peak of 1014 separate research tasks was slashed to just 364 in 2010. Space biology and the physical sciences were particularly hard hit, losing about 80% of their research portfolio.

Congress heard the research community's concerns, and we are most thankful for your response. The NRC's Life and Physical Sciences (LPS) Decadal Survey - completed in 2011 as a response to a request from Congress introduced in 2008 authorization language - prompted a sea change in NASA's approach to biological and physical sciences research.

The LPS Decadal summarized and sequenced 65 high priority research tasks. Furthermore, the Decadal study created two notional research plans aligned with specific priorities; one being a goal of rebuilding a research enterprise and the other a goal of a human mission to Mars. More about these goals later.

3. How Are Priorities for Research on the ISS Established and Is There a Clear and Well Understood Process for Aligning ISS Resources with Those Priorities?

My response to this question considers general aspects of peer-reviewed research projects that are solicited through open competition. All NASA-sponsored space life and physical sciences research is conducted in this way.

Developing strategic priorities for ISS research is not a new concept. Notable examples from this millennium include:

- The NASA-sponsored Research Maximization and Prioritization Task Force, commonly known as ReMAP, which reported its findings in 2002, representing the breadth of translational research in the biological and physical sciences.

- The ISS utilization studies organized by the National Research Council in 2005.
- Most recently, the Life and Physical Sciences (LPS) Decadal Research Plan; the first decadal survey of NASA's life and physical sciences programs. The guiding principle of the study was, "to set an agenda for research in the next decade that would use the unique characteristics of the space environment to address complex problems in the life and physical sciences, so as to deliver both new knowledge and practical benefits for humankind as it embarks on a new era of space exploration." Furthermore, the LPS Decadal organizers were tasked with establishing priorities for an integrated portfolio of biological and physical sciences research in the decade of 2010-2020.

Why have we asked the prioritization question so many times, and why must we do so again? Because space research informs two broad, often competing, goals: One centers on intrinsic scientific importance or impact; research that illuminates our place in the universe, but cannot be accomplished in a terrestrial environment. The other goal values research that enables long-term human exploration of space beyond low-earth orbit, and develops effective countermeasures to mitigate the potentially damaging effects of long-term exposure to the space environment. Over the past 25 years, other review panels, both internal and external to NASA, have defined similar goals. In the case of the LPS, research was categorized as either (1) required to enable exploration missions or (2) enabled or facilitated because of exploration missions. I prefer the more contemporary synonyms of "discovery" and "translational" research.

Throughout the history of the United States space program both goals have been important, but their relative importance has changed over time. In the early part of the Apollo era, the limited amount of biological and physical research that occurred was focused on the health and safety of astronaut crews in a microgravity environment. Until late in the Apollo program, significant research questions that did not contribute directly to a successful Moon landing received lower priority. In contrast, more regular access to space provided by the space shuttle afforded an opportunity for discovery research to take higher priority; an emphasis that fared poorly in the austere NASA budgetary environment of the mid-2000's.

Thus, the relative priority of these two goals of research - enabling long-term human exploration of space (translation) and answering questions of intrinsic scientific merit (discovery enabled by space research) – shifts according to NASA's programmatic goals.

I make note of the fact that section 201 NASA Authorization Act of 2015 articulates a translational goal of sending humans to Mars, while section 718 emphasizes discovery research. The key question is this: *Shall discovery or translational research takes precedence in the mature years of the ISS research program?* If it is translational research to prepare for a human trip to Mars, then the ISS research portfolio should be tailored accordingly.

The LPS Decadal Survey provides a very detailed scheme to evaluate the importance of proposed research on the International Space Station. It includes eight unique criteria to prioritize research[4], as follows:

- *Positive Impact on Exploration Efforts, Improved Access to Data or to Samples, Risk Reduction.* The extent to which the results of the research will reduce uncertainty about both the benefits and the risks of space exploration.
- *Potential to Enhance Mission Options or to Reduce Mission Costs.* The extent to which the results of the research will reduce the costs of space exploration.
- *Positive Impact on Exploration Efforts, Improved Access to Data or to Samples.* The extent to which the results of the research may lead to entirely new options for exploration missions.
- *Relative Impact Within a Research Field.* The extent to which the results of the research will provide full or partial answers to grand science challenges that the space environment provides a unique means to address.
- *Needs that are Unique to NASA Exploration Programs.* The extent to which the results of the research are uniquely needed by NASA, as opposed to any other agencies.
- *Research Programs That Could Be Dual-Use.* The extent to which the results of the research can be synergistic with other agencies' needs.
- *Research Value of Using Reduced-Gravity Environment.* The extent to which the research must use the space environment to achieve useful knowledge.
- *Ability to Translate Results to Terrestrial Needs.* The extent to which the results of the research could lead to either faster or better solutions to terrestrial problems or to terrestrial economic benefit.

Some of these criteria emphasize discovery; others translation. The LPS Decadal Survey prioritizes specific research tasks for each criterion. Again, the Survey appropriately stopped short of weighting or prioritizing criteria against each other because of the programmatic implications. That responsibility – to prioritize either discovery research or Mars - falls largely to the executive and legislative branches. When this question is decided, then the LPS decadal should be a useful tool to program research for the remaining life of the ISS.

Operationally, the ISS Program Office prioritizes all the research to be conducted on each ISS increment. It is a well understood process: CASIS receives a 50% allocation, followed by human research, then technology demonstrations. What resources remain are allocated to the Biological and Physical Sciences Program and the Science Mission Directorate payloads. Both the Human Research and Biological and Physical Science utilize the LPS Decadal criteria for prioritization within their respective programs, but it is not apparent the extent, if any, that LPS Decadal criteria are used to prioritize research across the four programs.

Lastly, it is worth noting that ISS research expenditures, which in FY 2012 constituted about 8%, or $225M, of ISS program costs, are not anticipated to keep pace with overall cost growth of the ISS program.

4. What Are the Implications of the Proposed Extension of ISS Operations to 2024 on Research and What Criteria Should Congress Use to Consider the Proposed Extension?

To evaluate the proposed extension, one of the first tests that Congress should apply can be answered with a yes or no. "Is NASA prepared to operate a robust research program through 2024?" In my opinion, the answer is an unqualified, "yes!" The scope of change in NASA life and physical sciences in the past four years has been remarkable. Allow me to highlight some notable examples:

- In 2011 NASA reorganized the remnants of a once robust life and physical sciences program to form the Space Life and Physical Sciences Research and Applications Division (SLPSRA). The program is formulated to execute high quality, high value research and application activities in the areas of space life sciences, physical sciences and human research. This reorganization acknowledges – in point of fact, celebrates – both the discovery and translational outcomes of research in the biological and physical sciences.

- Consistent with recommendations in the LPS Decadal, the Biological and Physical Sciences Program has restarted regular research announcements for ground-based and flight experiments. As a rule, these proposals are externally peer reviewed. In FY2014, 30 proposals were funded; 9 of them flight experiments.
- NASA is making greater use of advisors in the National Academies of Science, Engineering and Medicine. In October of 2014 the NRC instituted a new Committee on Biology and Physical Sciences in Space (CBPSS) chaired by Betsy Cantwell (University of Arizona) and Rob Ferl (University of Florida). Part of the Committee's charge is to monitor the progress in implementation of the recommendations contained in, the LPS Decadal.
- The Human Research Program has been aligned with a global exploration strategy. Annual solicitations for research have resumed. The past four quarters for which summaries are available included 212 research publications and more than 277 research proposals.
- We now have an American astronaut on a one-year mission to the ISS, with a unique opportunity to examine his genomic response to this environment.
- The technical content of the Human Bioastronautics Roadmap is in the middle of a five-year review of its 33 risks and 299 research gaps relevant to health and operations in space. The project is being conducted by the Institute of Medicine.
- NASA's Human System Risk Board tracks a subset of 23 risks that require additional research. While all but one have some level of risk mitigation for a one-year stay on the Moon, about half (N=11) do not have any substantive level of risk mitigation for three-year planetary operations.

I think it's reasonable to conclude that NASA has planned its life and physical sciences enterprise to take advantage of ISS research capabilities. The greatest remaining knowledge gaps are for Design Reference Missions on Mars for more than one year.

A recent NASA Office of the Inspector General (OIG) report[5] identified several concerns for continued ISS operations through 2024. There are four aspects of the report that I'd like to address:

First, the OIG found that ISS extension to 2024 could permit NASA enough time to mitigate an additional seven risks of long duration spaceflight. Nevertheless, extended utilization was not expected to fully mitigate another 11 human health risks prior to 2024, and two additional risks could not be mitigated using the ISS. The OIG concluded that NASA, "needs to prioritize its research aboard Station to address the most important risks in the time available." I think this conclusion misses an important point. The likelihood and consequences of at least 11 of the 13 unmitigated risks are dependent on the tasks required of a crew during a Mars Design Reference Mission. Today, there are simply too many degrees of freedom in the task set to establish useful risk criteria. Therefore, before the capabilities of the ISS to mitigate these risks can be evaluated, the risk must be better understood by performing a thorough task analysis of Martian operations.

Second, the report did not address powered down mass to any great extent. This is a critical need when biological samples, including live organisms, are to be returned to the ground for additional study.

Third, the OIG emphasized average crew time as a metric to quantify research utility. Although there are other metrics, including number of investigations, use of allocated space, up-mass, down-mass, and power, thermal, and data usage; in general, NASA does not consider these measures primary indicators of research utilization[6]. What is missing is a method to evaluate the *efficiency* of on-orbit research. Specifically, what percentage of crew time allocated to research is used to conduct it, compared to ancillary functions for such as setting up and stowing equipment? A similar focus has improved extravehicular operations on the ISS. I suspect that we will find that some of the highest priority research, such as studies using small mammals, is also the least efficient; requiring substantial amounts of crew time to set up experiments. If this is true, then increasing efficiency, for example, by improving coordination between NASA and CASIS, could be another way to capture more crew time for research in high priority areas.

Fourth, the OIG notes that research time is constrained with a six person crew. To maximize research utilization, we need to think about a seventh scientist crew member when commercial crew systems can support him or her.

Summary

We desperately need to increase research capabilities in space by translating findings from cell culture to reference organisms and mammalian models such as mice and rats to future flight crews. Translational research is the "gold standard" of the NIH, and it is what the research community, and the American people, should expect from the International Space Station. We need the capability to house and test model organisms on the ISS for extended periods of time, and whenever possible, to expose them to loading forces that approximate Mars. But equally important, we need adequate time for crew to prepare and conduct these experiments. The potential return is immense; the application of this research to our aging public could become one of the most important justifications for an extended human presence in space.

My LPS Decadal Survey colleagues and I contend that NASA can and should continue to restore a high level of programmatic vision and dedication to life and physical sciences research, to ensure that the considerable obstacles to human exploration missions to Mars can be resolved. This will depend on NASA embracing life and physical sciences research as part of its core exploration mission and re-energizing a community of life and physical scientists and engineers focused on both discovery and translational research.

To maximize ISS research, it is of paramount importance ...

- That the life and physical sciences research portfolio supported by NASA, both extramurally and intramurally, receive high attention.
- That NASA's research management structure be optimized to meet its discovery research, translational research, and commercialization goals. The utility of a coherent research plan that is appropriately resourced and consistently applied to enable exploration cannot be overemphasized. This will require improved coordination with CASIS.
- That the research portfolio be based on both discovery and translational programmatic priorities, and with specific destination(s) and mission tasks in mind.
- That there is sufficient external oversight to help NASA reach its research goals.

My top recommendations are the following:

- Articulate a timeframe for delivering and completing an operational risk mitigation plan for a multi-year human mission to Mars, and vet both the plan and the timeframe with the external scientific community.

- Review the essential resources for extended mammalian research on the ISS, including a seventh crew member; a scientist-astronaut whose nominal responsibilities are science programming.
- Extend biological science experiments to cover a substantial portion of a mammalian life cycle, and incorporate fractional (Martian) gravity exposure where possible.

Mr. Chairman, given sufficient resources, I am optimistic that NASA can deliver another decade of rigorous translational research. It's what the scientific community expects, and the American people deserve. I sincerely thank you for your vigilant support of the nation's space program, and the opportunity to appear before you today.

End Notes

[1] http://www.nap.edu/catalog/13048/recapturing-a-future-for-space-exploration-life-and-physical-sciences.

[2] http://www.nasa.gov/pdf/396093main_HSF_Cmte_FinalReport.pdf.

[3] http://www.gutenberg.org/ebooks/15000.

[4] http://www.nap.edu/catalog/13048/recapturing-a-future-for-space-exploration-life-and-physical-sciences.

[5] http://oig.nasa.gov/audits/reports/FY14/IG-14-031.pdf.

[6] https://oig.nasa.gov/audits/reports/FY13/IG-13-019.pdf.

In: The International Space Station
Editor: Evelyn Clemens
ISBN: 978-1-63484-071-2

Chapter 7

INTERNATIONAL SPACE STATION: APPROACHES FOR ENSURING UTILIZATION THROUGH 2020 ARE REASONABLE BUT SHOULD BE REVISITED AS NASA GAINS MORE KNOWLEDGE OF ON-ORBIT PERFORMANCE*

United States Government Accountability Office

This report was revised on March 19, 2012.

WHY GAO DID THIS STUDY

In 2010 the National Aeronautics and Space Administration (NASA) was authorized to extend the life of the International Space Station (ISS) from 2015 through at least September 30, 2020. Gauging the feasibility of doing so is quite complex. Among the factors to be assessed are the reliability of key components, NASA's ability to deliver spares to the ISS, the projected life of structures that cannot be replaced, and the health of systems that affect safety.

---

* This is an edited, reformatted and augmented version of a United States Government Accountability Office publication, No. GAO-12-162, dated December 2011.

While some empirical data exist, the ISS is a unique facility in space and assessing its extended life requires the use of sophisticated analytical techniques and judgments. GAO provided a preliminary report on NASA's use of such techniques in April 2011. For this review, GAO assessed the extent to which NASA has ensured essential spare parts are available and ISS structures and hardware are sound for continued ISS utilization through 2020. GAO interviewed NASA officials and outside experts; assessed the methodology underlying NASA's findings; conducted a limited test of data supporting NASA's assessments; and analyzed documentation such as ongoing assessments, schedules and other relevant efforts.

What GAO Recommends

GAO recommends that NASA reassess the relative weight given to original reliability estimates of spares' life expectancies as performance data accumulates and document the rationale behind specific performance goal decisions. NASA concurred with our recommendations.

What GAO Found

NASA's approach to determining, obtaining, and delivering necessary spare parts to the ISS is reasonable to ensure continued utilization of the station through 2020. The statistical process and methodology being used to determine the expected lifetimes of replacement units is a sound and commonly accepted approach within the risk assessment community that considers both manufacturers' predictions and the systems' actual performance. To date NASA has given equal weight to manufacturers' predictions and actual performance, and currently has no plans to reassess this decision. However, as time goes on, the resulting estimates could prove to be overly conservative, given that NASA has found failure rates for replacement units to be lower than manufacturers' predictions. Therefore, continuing to weigh the manufacturers' predictions equally with actual performance could lead NASA to purchase an excess of spares. NASA also has a reasonable process for establishing performance goals for various functions necessary for utilization and determining whether available spares are sufficient to meet goals through 2020, but the rationale supporting these decisions has not been systematically documented.

NASA is also using reasonable analytical tools to assess structural health and determine whether ISS hardware can operate safely through 2020. On the basis of prior analysis of structural life usage through 2015 and the robust design of the ISS structures, NASA currently anticipates that—with some mitigation—the ISS will remain structurally sound for continued operations through 2020. NASA also is using reasonable methodologies, governed by agency directives and informed by NASA program experts, to assess safe operations of the ISS as a whole as well as identify replacement units and other hardware, failure of which could result in an increased risk of loss of station or loss of crew through 2020. NASA plans to develop, through 2015, methods to mitigate issues identified and expects to begin implementing corrective actions as plans are put in place.

ABBREVIATIONS

ATV	Automated Transfer Vehicle
BSP	baseband signal processor
COTS	Commercial Orbital Transportation Services
DOD	Department of Defense
ECLSS	Environmental Control and Life Support System
FGB	Functional Cargo Block
HTV	H-II Transfer Vehicle
ISS	International Space Station
MTBF	mean time between failure
MADS	Modeling and Analysis Data Set
NASA	National Aeronautics and Space Administration
ORU	orbital replacement unit
Orbital	Orbital Sciences Corporation
PRA	Probabilistic Risk Assessment
SASA	S-Band antenna and support assembly
SpaceX	Space Exploration Technologies
TDRSS	Tracking and Data Relay Satellite System

December 15, 2011

The Honorable John D. Rockefeller IV
Chairman
The Honorable Kay Bailey Hutchinson
Ranking Member
Committee on Commerce, Science, and Transportation
United States Senate

The Honorable Ralph Hall
Chairman
The Honorable Eddie Bernice Johnson
Ranking Member
Committee on Science, Space and Technology
House of Representatives

From 1994 through 2010, the National Aeronautics and Space Administration (NASA) estimates it directly invested over $48 billion in development and construction of the on-orbit scientific laboratory, the International Space Station (ISS). To take advantage of that investment, the NASA Authorization Act of 2010 extended the mission of the ISS by a minimum of 5 years, from 2015 through at least September 30, 2020. Traditionally, the ISS received the majority of its supplies and spare parts via the space shuttle. The space shuttle conducted its final flight in July 2011 and as a result NASA now plans to rely on a combination of foreign and commercially developed domestic launch vehicles to supply hardware spares and supplies to the ISS through 2020.

In consideration of the shuttle's retirement, section 503 of the NASA Authorization Act of 2010, directed NASA to assess its plans for ensuring essential spares were available to support the ISS through its planned operational life, and to report to Congress on that assessment.[1] The same law also directed that GAO provide an evaluation of the accuracy and level of confidence in the findings contained in NASA's assessment of the essential modules, operational systems and components, structural elements, and permanent scientific equipment required to ensure complete, effective and safe functioning and full scientific utilization of the ISS through 2020.

Gauging the feasibility of extending the life of the ISS is no simple exercise. Among the many factors to be assessed are the reliability of key components, NASA's ability to deliver spares to the ISS, the projected life of structures that cannot be replaced, and in-depth analysis of those components and systems that affect safety. While some empirical data exists, because the ISS is a unique facility in space, assessing its extended life necessarily requires the use of sophisticated analytical techniques and judgments.

NASA reported to Congress on January 11, 2011, that the ISS could be effectively maintained through 2020 within the normal range of human spaceflight risk with a combination of existing and planned re-supply vehicles. Subsequently, we issued our initial evaluation on April 11, 2011.[2] We found that NASA was using analytical techniques, physical tests, and inspections to assess ISS structures and hardware as well as to determine spare parts needed to support safe functioning and full utilization of the ISS through 2020. Additionally, we noted that NASA's assessments appeared to be supported by sufficient, accurate, and relevant underlying data. We found, however, that NASA's estimates of ISS sparing needs were sensitive to assumptions about hardware reliability.

As we noted in our report, these assessments were ongoing, so not all results were available. At your request, we continued our evaluation of NASA's plans to ensure the ISS is supported through 2020. Specifically, we assessed (1) the extent to which NASA has ensured spare components essential for ISS operations will be available through 2020, and (2) the extent to which NASA has assurance that the ISS structure and hardware will be sound through 2020 and potentially beyond. To conduct this work, we examined NASA's processes for quantifying ISS maintenance needs. We interviewed knowledgeable officials and experts within and outside of NASA to evaluate and obtain an understanding of the methodologies and analytic techniques NASA and other organizations use in addressing supportability. We conducted a limited test of the reliability of key data supporting NASA's assessments. We also examined the schedules for and manifests of planned re-supply missions and evaluated the status of ongoing efforts to estimate the costs of required spares. We obtained and analyzed available results of ongoing assessments of structural health and hardware systems. See appendix I for our detailed scope and methodology.

We conducted this performance audit from February 2011 to November 2011 in accordance with generally accepted government auditing standards. Those standards require that we plan and perform the audit to obtain sufficient, appropriate evidence to provide a reasonable basis for our findings and conclusions based on our audit objectives. We believe that the evidence obtained provides a reasonable basis for our findings based on our audit objectives.

BACKGROUND

The ISS is composed of about 1 million pounds of hardware, brought to orbit over the course of a decade. It is the largest orbiting man-made object and was constructed to support three activities: scientific research, technology development, and development of industrial applications. The facilities aboard the ISS allow for ongoing research in microgravity, studies of other aspects of the space environment, tests of new technology, and long-term space operations. The facilities also enable a permanent crew of up to six astronauts to maintain their physical health standards while conducting many different types of research, including experiments in biotechnology, combustion science, fluid physics, and materials science, on behalf of ground-based researchers. Furthermore, the ISS has the capability to support research on materials and other technologies to see how they react in the space environment. In addition to these research activities, the ISS crew conducts preventive and corrective maintenance to the on-board systems.

The ISS includes (1) primary structures, i.e., the external trusses which serve as the backbone of the station and the pressurized modules that are occupied by the ISS crew, and (2) functional systems comprised of orbital replacement units (ORU), i.e., systems that provide basic functionality such as life support and electrical power which are comprised of modular components that are easily replaced by astronauts on orbit.

NASA originally planned for an ISS lifespan of 15 years—5 years to assemble the individual components of the ISS on-orbit and 10 years for operations. To ensure structural soundness for 15 years, NASA required ISS developers to analytically demonstrate that ISS primary structures would remain structurally sound for four service lives; that is, at least 60 years. This demonstration involved showing analytically that the largest undetected crack that could exist in the structure would not grow and cause failure of the structure in four service lifetimes[3] when subjected to the mechanical, thermal, and pressure loads[4] of the on-orbit environment.

Although the ISS originally was planned to operate until 2015, the NASA Authorization Act of 2008 directed the agency to do nothing to preclude the use of the ISS beyond 2015 and the NASA Authorization Act of 2010 directed the agency to extend operations through at least September 30, 2020.[5] NASA had originally planned for the space shuttle to serve as the means of transporting hardware and supplies to the ISS through the end of the ISS's life. However, in 2004, President George W. Bush announced his Vision for Space Exploration that included direction for NASA to pursue commercial opportunities for providing transportation and other services to support the ISS after 2010.[6] As a result of that direction, in July 2011, the space shuttle flew its last mission to the ISS.

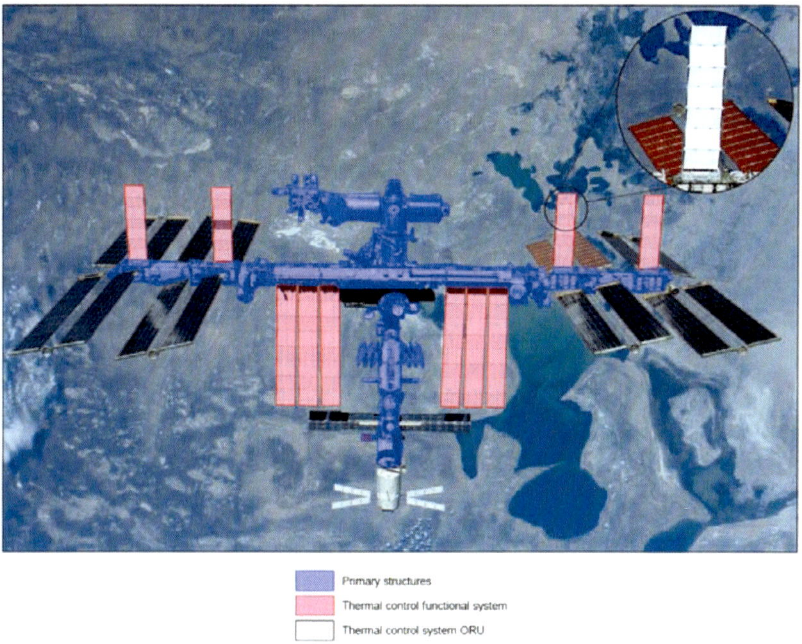

Source: NASA.

Figure 1. ISS Primary Structures, Thermal Control Functional System, and ORU.

NASA's current strategy for supporting the ISS relies on a mixed fleet of vehicles, including those developed by ISS international partners and commercial domestic launch providers, to service the ISS through 2020. International partner vehicles include the European Space Agency's Automated Transfer Vehicle (ATV), and the Japan Aerospace Exploration Agency's H-II Transfer Vehicle (HTV). NASA has agreements in place with the respective European and Japanese consortiums for the ATV and HTV to provide supplies and spares to the ISS. The domestic commercial launch vehicles, fostered under a NASA-initiated project known as Commercial Orbital Transportation Services (COTS), are being developed by private industry corporations Space Exploration Technologies (SpaceX) and Orbital Sciences Corporation (Orbital). In 2006 and 2008, NASA awarded agreements to SpaceX and Orbital for development and demonstration of cargo transport capabilities, which SpaceX and Orbital are providing through their respective Falcon 9 and Taurus II launch vehicles. In late 2008, NASA awarded both companies commercial resupply services contracts to provide supplies and spares to the ISS through 2015. NASA plans to use the SpaceX and Orbital Sciences launch vehicles to supply the ISS beginning in 2012 and continuing through 2020.

Initial ISS Supportability Strategy

NASA's strategy for supporting the ISS at the beginning of the program was based on several key factors, including:

- a specified level of ORU design reliability;
- an ISS design with electrical and mechanical functions modularized into ORUs for simple on-orbit removal and replacement;
- a consistent, reliable transportation service to and from the ISS in the form of the space shuttle; and
- an approach that used the space shuttle as the vehicle for bringing failed ORUs to earth and flying repaired ORUs back to the ISS.

The ISS was designed and certified based on the idea that its 15-year design life would be achieved through returning failed or degraded ORUs to earth for repair. NASA established ISS initial estimates of spare ORUs needed using the equipment manufacturers' original reliability predictions, and a transportation plan that assumed space shuttle operations would continue until the planned end of ISS life in 2015. This strategy assumed not only that the space shuttle would provide a reliable means of delivering spares to the ISS, but also a means of returning failed or expended hardware to earth. NASA retained many of the original equipment manufacturers of ISS hardware to provide sustaining engineering and repair or refurbishment capabilities for the planned life of the ISS.

Historically, NASA's methods for determining the required number of ISS spares used a combination of the original equipment manufacturers' projections about the length of time an ORU would operate before randomly failing and the professional judgment of NASA engineers. NASA and the original equipment manufacturers express an ORU's expected frequency of random failure as the mean time between failure (MTBF), which predicts, in hours, the average amount of time an ORU will operate before a random failure occurs. NASA considers the MTBF as the primary predictive measure of an ORU's reliability and uses this information to predict when ORUs will require maintenance to prevent or correct failures. To calculate an MTBF, the ORU's original equipment manufacturer would typically use reliability models that were populated with the failure rates of the individual parts obtained from reliability handbooks, such as Military Handbook 217F and other authoritative sources.[7] These models also incorporate data on part quality and quantity as well as assumptions about the ORU's operating temperature, operating environment, and the amount of time it spends in an active state. According to agency officials, the MTBF calculated by using this process would typically be provided to NASA in a reliability report as part of the technical data package accompanying the delivery of the ORU. Officials indicated for example, as part of the technical data package accompanying the delivery of the ISS ammonia/water heat exchanger, which helps regulate heat on the ISS, Boeing prepared a reliability verification analysis report that estimated the MTBF for this ORU is 362,401 hours.

Revised ISS Supportability Strategy

NASA has revised its ISS supportability strategy for ORUs from the original "repair and return" strategy to one of "build and dispose." Three factors played major parts in this revision:

- the decision to decommission the space shuttle;
- NASA's observation that reliability of on-orbit ORUs has been much higher, on average, than the original equipment manufacturers' initial predictions of MTBF; and
- NASA's recognition that procurement of enough spares to ensure full functionality through the end of ISS life, based on manufacturers' predictions, would be unaffordable and was not necessary to ensure safe, useful ISS operations.

Lacking a means upon retirement of the space shuttle to return failed ORUs to earth, repair them, and take them back to the ISS, NASA moved to a strategy of replacing failed ORUs with new ones. NASA pre-positioned spares on-board the ISS to ensure that the station could operate in the intervening period between shuttle retirement and expected development of new launch vehicles. Because NASA no longer has the sparing flexibility provided by the space shuttle, and astronauts can perform only minor repairs while on-orbit, NASA's current supportability strategy relies more heavily on modeling and computational analysis to determine the quantity of spare ORUs needed to support the station.

NASA also revised its ISS supportability strategy because ORU on-orbit failure data indicated that the manufacturers' MTBFs were, on average, too conservative. According to NASA representatives, this conservatism resulted from original equipment manufacturers' tendency to err on the side of caution, that is, the manufacturers were primarily interested in demonstrating through analysis that ORUs they developed would meet minimum MTBF criteria established by NASA. Moreover, the component-level reliability data contained in Military Handbook 217F that served as the basis for original equipment manufacturers' analysis has also proved to be overly conservative in many cases.

Given the actual on-orbit failure data, NASA recognized that a sparing strategy built around manufacturers' original MTBF estimates needed to be modified or it would result in the acquisition of many more spares than were truly needed to support the program. Consequently, the ISS program set up a team of experts referred to as a "tiger team" to evaluate other methods of determining more accurate MTBF estimates for ORUs. In 2005, this team settled on a new NASA process that used a Bayesian estimation process to update the ORUs' original MTBFs. The Bayesian estimation process combines the manufacturer's original MTBF with actual on-orbit data—including the accumulated hours of operation and random failure history of an ORU population—and results in an updated MTBF reflecting NASA's experience with ORU performance.[8] In January 2006, the ISS Reliability and Maintainability Panel approved the use of the Bayesian estimation process to update MTBF estimates.[9] Since that time, NASA has implemented an annual Bayesian MTBF update process that continues to refine reliability and failure projections for ISS ORUs.

While an improved method for updating the MTBF of each ORU was needed to inform NASA's sparing decisions, it was not sufficient by itself to predict the required quantity of spares needed, since the reliability of individual ORU is only one piece of the puzzle. Groups of ORUs are arranged into an ORU architecture to support a specific function. Many functional architectures include redundancy within their design so that the failure of a single ORU does not disable the function. For example, the S-band communications and tracking function—which supports all command and control from the ground—consists of six ORUs arranged into two independent strings. 10 Each string consists of three ORUs: (1) the S-Band antenna and support assembly (SASA); (2) the Tracking and Data Relay Satellite System (TDRSS) Transponder; and (3) the baseband signal processor (BSP). Each string has the capacity of being designated prime or backup using a software switch which automatically turns on the backup if the prime fails. Figure 2 illustrates the ORU functional architecture for the S-band communications and tracking function.

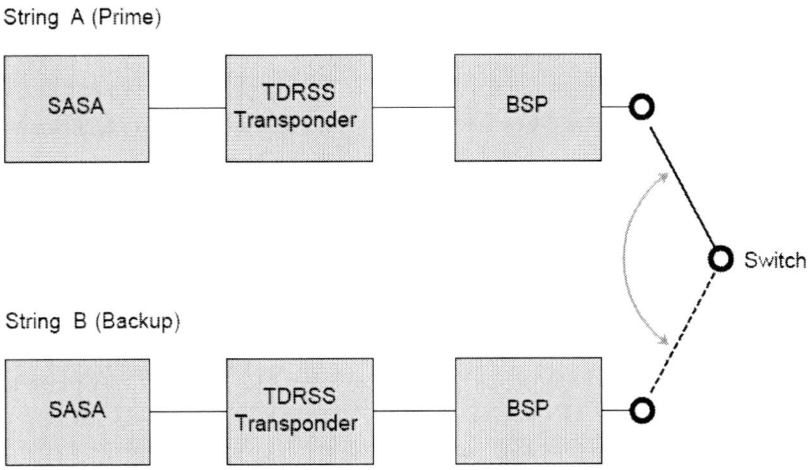

Source: GAO analysis of NASA data.

Figure 2. S-Band Functional Architecture.

A failure of a single ORU will disable its string, but will not disable the function since the function can be switched over to the second string.

To enable the ISS program to predict how many spare ORUs it required, NASA needed not only the best available estimate of each ORU's MTBF, but also a means of quantifying what level of redundancy was required for each function, and with what level of confidence each function must be maintained. When it considered the challenges of supporting ISS without the space shuttle's ability to guarantee the repair and refurbishment of failed ORUs, the ISS program recognized that the ISS would need to "gracefully degrade" without compromising safety as it approached its end of life. Graceful degradation was defined by the ISS program for 27 functions through the establishment of a functionality target at the end of the ISS's life, and a confidence goal for meeting that functionality target. The ISS program also recognized that it would need an appropriate analytic method for predicting whether it would meet its functionality targets and confidence goals. The functionality target and confidence goal for each function drives the quantity of spare ORUs required through the end of the ISS's life. For example, the ISS program established a functionality target for its S-band communications and track function of having at least one of the two ORU strings operating at end of life, and a 98 percent confidence goal for meeting this functionality target. The ISS program has conducted an annual functional availability risk assessment of 27 functions every year since 2006 to assess whether or not the ISS is meeting its functionality targets and confidence goals through 2015. In 2010 NASA conducted the first assessment addressing ISS functionality through 2020. This assessment employs a Monte Carlo model to calculate, for each ISS function, the probability of meeting its functionality targets as a function of time through end-of-life under different sparing quantity assumptions.[11] The assessment uses the results from the Bayesian MTBF process as an input.

NASA's Sparing Approach to Ensure Continued ISS Operations and Utilization through 2020 Employs Appropriate and Reasonable Methodologies

NASA has chosen appropriate and reasonable methodologies to predict the quantity of spares it requires to maintain ISS operations and full scientific utilization through 2020. NASA uses a Bayesian estimation process to improve estimates of ORU MTBF and functional availability risk assessments to determine with what confidence ISS functionality targets can be maintained. The Bayesian estimation process is an appropriate and reasonable technique for updating the original predicted MTBF of ISS ORUs. Similar techniques have been used on other NASA programs and are used throughout government and industry to update reliability estimates for nuclear power plant components, pressure vessels at oil and gas facilities, and medical devices. Moreover, the ISS program's specific implementation of this process is reasonable, given its current level of knowledge of the on-orbit performance of ORUs. However, the results of the Bayesian estimation process are sensitive to key assumptions, particularly with regard to the weight given to the manufacturer's original estimated MTBF relative to the actual on-orbit performance and failure data. The ISS program's functional availability risk assessment process is also an appropriate and reasonable technique for predicting whether available spares are sufficient to allow the ISS to meet its functionality targets and confidence goals, set by the program manager with input from experts. Moreover, the ISS program conducts sensitivity analysis to determine how the number of required spares would change if the functionality targets and confidence goals were relaxed. The ISS program manager's rationale for selecting these targets and goals, however, has not been systematically documented.

Bayesian Estimation Process Is an Appropriate and Reasonable Technique for Updating Estimates

The ISS program's Bayesian estimation process employs a common statistical technique for updating the original reliability estimates of ISS ORUs. The process combines the probability distribution of the original equipment manufacturer's estimated MTBF with actual on-orbit data, which includes the accumulated hours of operation and random-failure history of the ORU population.[12] The result is a probability distribution of a new "operational MTBF" reflecting NASA's experience with the ORU. The basic idea underlying Bayesian estimation is that an original estimate for a statistical parameter—the MTBF, for example—can be improved upon by combining it with actual experience. As more experience is gained and applied in the updating process, the accuracy of the new distribution that defines the MTBF will improve. If the original distribution has a relatively large variance—for example, substantial uncertainty in the value of the MTBF—then the new evidence (accumulated hours of operation and the number of on-orbit random failures) will tend to have a larger effect on the new distribution. On the other hand, if the original distribution has a relatively small variance—that is, a high degree of certainty that the value of the MTBF falls between a relatively narrow range—then the new evidence will have less influence on the new distribution. Figure 3 illustrates a typical example of the relationship between the original and updated probability distributions.

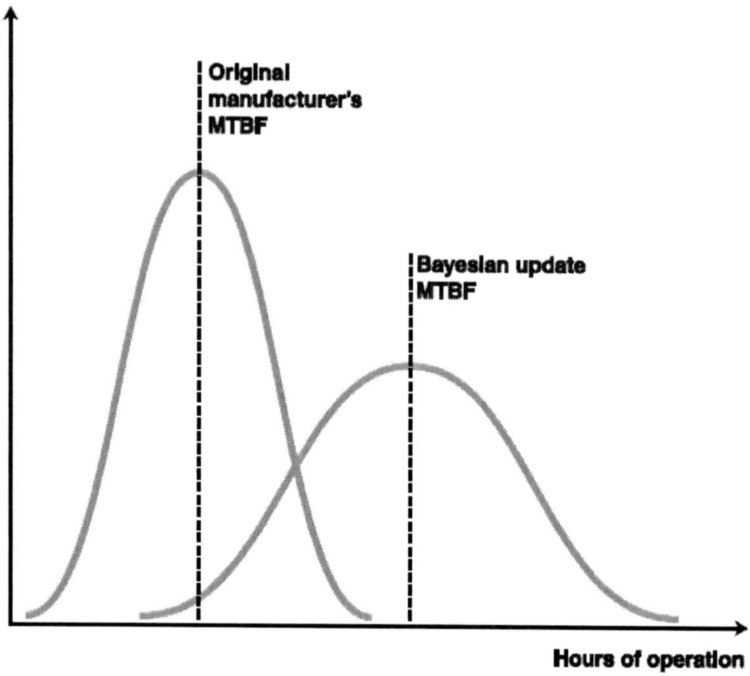

Source: GAO analysis of NASA data.

Figure 3. Hypothetical Example of the Relationship between Original and Updated Probability Distributions.

According to NASA representatives, the Bayesian estimation process was found to be the most appropriate for ISS given:

- the Bayesian process is an accepted method to update data within the probabilistic risk assessment discipline;
- the process incorporates original reliability data with observed performance data; and
- the Bayesian process can be performed even with no observed random failures, which is important because many types of ORU have experienced no failures over the life of the ISS.[13]

Statistics textbooks and other authoritative sources of guidance describe the use of the Bayesian estimation process to update statistical variables, including MTBF. Moreover, the Bayesian estimation process is a key element of NASA's recommended approach for conducting probabilistic risk assessments, a basic requirement for all major NASA programs and projects.[14] NASA's probabilistic risk assessment guidance describes the importance of the Bayesian estimation process, and provides broad implementing guidance for NASA program managers on incorporating Bayesian estimation into these assessments. According to representatives from NASA's Office of Safety and Mission Assurance — the organization within NASA that developed NASA's probabilistic risk assessment guidance—the use of Bayesian estimation is critical to ensuring that a program's probabilistic risk assessment is supported by the best available evidence of system reliability and performance. NASA officials indicated that Bayesian estimation has proven to be more accurate than original manufacturers' reliability estimates in predicting overall failure rates for ORUs on the ISS. Bayesian estimation has been used to update MTBFs for components on numerous NASA programs, including the space shuttle, the Hubble Space Telescope, and the TDRSS family of satellites.

Similar methodologies are also used within government and industry to update values for statistical parameters:

- The Nuclear Regulatory Commission employs the Bayesian estimation process to update the failure rates of components in nuclear power plants.
- The oil and gas industry employs the Bayesian estimation process to update the failure rates of pressure vessels and other types of equipment at industrial oil and gas facilities.
- The Food and Drug Administration has developed guidance for the assessment of medical device effectiveness based on the Bayesian estimation process.

While NASA's Office of Safety and Mission Assurance provides NASA programs with broad implementing guidance on employing the Bayesian estimation process, it leaves the details of implementing this process to the discretion of the individual programs. The ISS program's application of the Bayesian estimation process requires four key sources of input data: (1) an ORU's original manufacturer-estimated MTBF; (2) a parameter that governs the variance of the original MTBF's probability distribution—a measure of the dispersion of values about the MTBF's most likely value; (3) the total aggregated hours of operation for the ORU population under consideration;[15] and (4) the actual number of random failures experienced by the population of ORUs.

For three of the four sources of input data, the ISS program generally has complete data readily available for each set of ORUs. For example, each ORU's original estimated MTBF was obtained from the reliability analysis performed by the original equipment manufacturer and incorporated into an authoritative ISS database.[16] In addition, detailed data on each ORU's total hours of operation and the number of random failures is maintained by the ISS program's logistics and maintenance team, and is therefore readily available as a source of reliable input data.

For the fourth source of input data—the parameter that governs the variance of the original MTBF's probability distribution—there is no readily available source. In most cases, originally predicted MTBF estimates were provided to the ISS program from original equipment manufacturers as a single value, without any estimate of variance. Therefore, an assumption about the variance of MTBF must be made. However, selection of a parameter that defines the variance of the original MTBF's probability distribution is the most challenging aspect of applying the Bayesian estimation process. The ISS tiger team consulted at least two sources of information including a 1990 study of external maintenance activities for Space Station Freedom—the precursor to ISS—which had recommended using a parameter that produced a relatively large variance for the original estimated MTBF, and a 2005 independent technical consulting report by NASA's Engineering and Safety Center (NESC), which recommended using a parameter that would produce a relatively small variance.[17] After deliberating, the tiger team recommended adopting a variance parameter for the original MTBF that would grant approximately "equal weight" to an ORU's original estimated MTBF and its on-orbit experience. The ISS program applies the Bayesian process to all ORUs that have experienced random failures. For an ORU that has not experienced any random failures, the ISS program allows each set of ORUs to operate for at least one-half of the number of hours that is equal to the original predicted MTBF for that set of units before updating the MTBF by applying the Bayesian estimation process.

The ISS program believes that its choice of variance parameter is appropriate and consistent with on-orbit experience to date. Using a parameter that produces a larger variance would place more weight on on-orbit experience. Because many ORUs have experienced no failures, however, this could lead to overly optimistic updates of MTBF. On the other hand, using a parameter that produces a smaller variance would place more weight on the original estimated MTBF and lead to continued unwarranted conservatism, according to the ISS program. Given the current state of knowledge about

ORUs, NASA's choice of an MTBF variance parameter appears to be reasonable and based on sound professional judgment. Thus far, NASA has not revisited the relative weight given manufacturers' original reliability estimates versus on-orbit experience, and currently has indicated no plans to do so. However, as time goes on, the variance parameter selected could prove to be overly conservative given that NASA has been seeing failure rates for ORUs that are far more in line with Bayesian updates than manufacturers' predictions. Continuing to weigh manufacturers' original estimates equally with on-orbit experience through the ISS end of life for those ORUs that have experienced a random failure could lead to the purchase of an excess of spares because NASA could continue to predict more failures than actually occur.

ISS program representatives have noted that the MTBFs based on implementation of their Bayesian estimation process—with their choice of MTBF variance parameter—are much closer to actual overall ORU failure rates. For example, in 2005, a comparison of actual and predicted failures of the external ISS ORU population showed the hardware experiencing only about one-third the predicted number of failures based on original MTBF estimates. However, given its selection of a parameter for MTBF variance, the operational MTBF calculated using ISS's Bayesian estimation process is sensitive to single failures—that is, a single failure of an ORU can dramatically change the operational MTBF calculated by using this process. In one sense, this Bayesian update process does exactly what one would expect—modify the original MTBF based on observed failure data. Because many ISS ORUs have operated for well beyond their original predicted MTBF without experiencing a random failure, and consequently have updated operational MTBFs that are several times greater than their original predicted MTBFs, a single random failure will quickly reverse this growth. For example, successive annual updates of the operational MTBF (in hours) for the ISS pump module assembly increased each year through 2009. A single failure of the pump module assembly in July 2010,[18] however, resulted in a significant decrease in MTBF—from an October 2009 estimate of 104,683 hours to an August 2010 estimate of 58,119 hours. This decrease of 44 percent could result in the need to procure additional spares. Table 1 shows how the estimated MTBF of the pump module assembly has changed over time.

Table 1. Successive Annual Updates of MTBF (in Hours) for the ISS Pump Module Assembly

Pump module assembly	Original MTBF in October 2006 data	Updated MTBF October 2007	Updated MTBF October 2008	Current MTBF October 2009	Estimated runtime hours at failure per ISS risk team to August 3, 2010	New MTBFBayesianupdated with onefailure August 2010
Hours	39,800	52,450	86,560	104,683	63,600	58,119

Source: NASA.

Similarly, the ISS television camera interface converter has not yet experienced any random failures. Consequently, its predicted MTBF has grown by more than 170 percent—from an original value of 100,675 hours to 272,908 hours (as of October 2010). A single random failure of this device today, however, would cut the value of the operational MTBF by more than half its current value—to about 134,000 hours. These large changes in MTBF may have consequences for the results of the ISS functional availability risk assessment, and could result in the need to procure additional spares as discussed below. Given the fact that the majority of ORUs on the ISS have yet to experience a random failure, this issue is one that could affect sparing decisions and place the program at risk of procuring too few spares. NASA officials indicated that the agency mitigates this risk by updating its sparing estimates annually.

Process for Establishing Functionality Targets and Confidence Goals Is Reasonable but Rationale for Decisions Is Not Well Documented

NASA has reasonable processes in place for establishing functionality targets and confidence goals for each ISS function being modeled in its functional availability risk assessment.[19] The ISS program manager considers the recommendations of system experts who validate each functionality target to ensure that it meets all science utilization and crew habitability requirements. The ISS program manager's rationale for establishing functionality targets and confidence goals is also driven by an assessment of the operational needs of ISS as a whole, and by whether the needs satisfied by a given function might be accomplished in some other way. NASA officials indicated, for example, that the program manager's decision to set the functionality target for S-band communications and tracking as "at least one of two strings operational" instead of "both strings operational" was influenced by the planned upgrades to ISS Ku-band capability, which will be able to perform some functions that could previously be performed only by using S-band, thereby mitigating the loss of S-band redundancy. Finally, ISS functionality targets and confidence goals are revisited based upon the results of each annual functional risk assessment. Table 2 below shows several examples of the 27 ISS functions NASA includes in its functional availability risk assessment.

Table 2. Examples of ISS Functions Included in a Functional Availability Risk Assessment

Function	Purpose
Command & Data Handling	Provides command and control capability of ISS via a network of computers, data buses, and software.
Electrical Power System	Generates, stores, and distributes electrical power.
Structures & Mechanisms	Provides the primary and secondary structures of the ISS, such as trusses and payload racks, as well as ISSmechanisms, such as the joints and gimbals for positioning solar arrays.
Thermal Control System	Maintains temperature control of the ISS.

Source: NASA.

Once initial functionality targets and confidence goals have been established for each of the 27 functions, the ISS program employs a Monte Carlo simulation to predict how many spare ORUs would be required to meet these targets and confidence goals through the ISS end of life. The ISS program also conducts a sensitivity analysis to determine how the number of required spares would change if the functionality targets and confidence goals were relaxed. For example, the sensitivity analysis may reveal that a 98 percent confidence goal for a given functionality target can be met with a given quantity of spares, while a 99 percent confidence goal would require two to three times as many spares. Similarly, analysis may reveal that while one functionality target (e.g., three of four strings operational) may be achievable with a certain number of spares, achieving a more demanding functionality target (e.g., four out of four strings operational) might require several times as many spares. This result would not necessarily support the less demanding functionality target and confidence goal in all cases—despite the significant increase in the number of spares required—since the 99 percent confidence goal may be justified based on safety-of-life concerns. This assessment, however, allows the ISS program to see the costs, in terms of the required number of spares, of setting functionality targets and confidence goals at particular levels.

On the basis of this analysis, a technical community of system experts supporting each modeled function validates each functionality target and confidence goal and recommends them to the ISS program manager. The ISS program manager is free to reject these recommendations based on his or her broader knowledge of all ISS functions and other factors affecting sparing decisions. For example, during the 2010 functional availability risk assessment,[20] the technical community for the S-band communications and tracking function recommended a functionality target of at least one of two strings operational, and a confidence goal of 99.5 percent for maintaining this target. The program manager agreed with the functionality target, but decided to set the confidence goal at 98 percent. By setting a lower confidence goal, the number of additional spares required dropped from a total of six for all three types of ORUs to just two.

According to the ISS program manager and the system experts who make the functionality target recommendations, they go through a deliberative process wherein options for functionality targets and confidence goals are developed and presented to the ISS program manager at the Space Station Program Control Board meeting. These presentations contain the technical community's recommendation on the most appropriate functionality target and confidence level for a given function. As illustrated above, the program manager has the discretion and authority to select functionality targets and confidence levels other than the ones recommended by the technical community. Agency officials indicated that while the program manager's rationale for not accepting the recommendation of the technical community is discussed at these meetings, the meeting minutes do not typically capture the rationale for the program manager's decision to deviate from the recommendation of the system experts. Our *Standards for Internal Control in the Federal Government*[21] require, however, that important events and transactions are clearly documented and readily available for examination. This type of documentation provides evidence for the benefit of future decision-makers to continue to ensure that the rationale remains valid.

NASA's ISS Sparing Plan Relies on International and Domestic Vehicles and Currently Includes Some Cargo Shortfalls

NASA plans to use both international and domestic commercial launch vehicles to ensure ISS spares are available through 2020 because these are NASA's only near-term options for resupplying the ISS. Planned launches, however, may not provide enough cargo capability to cover anticipated growth in scientific utilization or margin for unforeseen ISS maintenance needs, especially from 2015 through 2020. Additionally, delays in domestic commercial launch vehicle development could affect sparing deliveries planned for 2012 and beyond.

NASA plans to rely on ISS international partner and new commercial launch vehicles to meet crew needs as well as support maintenance and scientific experiments through 2020. The agency has agreements in place with the European and Japanese space agencies for their respective vehicles, the ATV and HTV, to conduct sparing missions beginning in 2012. The ATV and HTV are unmanned vehicles, which have flown to the ISS, and carry such items as hardware and water.[22] NASA expects to employ a total of 13 international partner launches—7 from 2012 to 2015 and 6 from 2016 through 2020. As we have reported, the international partner vehicles alone cannot meet ISS sparing needs. 23 Consequently, NASA plans to use two types of domestic commercial launch vehicles to maintain ISS from 2012 through 2020. Development of these vehicles—the Falcon 9 and Taurus II—was fostered under a NASA-initiated effort known as Commercial Orbital Transportation Services (COTS). These vehicles are being developed by private industry corporations SpaceX (Falcon 9) and Orbital Sciences (Taurus II). In late 2008, NASA awarded contracts to both companies to provide cargo transport services to the ISS. Only SpaceX's Falcon 9 vehicle is designed to safely return significant amounts of cargo to earth, such as scientific experiments. NASA anticipates that SpaceX will begin providing that capability in 2012.

Commercial vehicles are essential to sustaining and utilizing the ISS. As Table 3 below indicates, SpaceX and Orbital are scheduled to fly 19 (73 percent) of the 26 launches NASA plans through 2015 and follow-on SpaceX and Orbital commercial vehicles are expected to fly 24 (80 percent) of the 30 launches from 2016 through 2020.[24] Additional international partner flights from 2014 to 2020 beyond those already scheduled are an unlikely option, according to NASA officials, because the European and Japanese space consortium's manufacturing facilities are not equipped to manufacture vehicles at a rate faster than their current commitments.

Although NASA expects domestic commercial launch vehicles to deliver the bulk of cargo required by the ISS through 2020, NASA strategic planning manifests indicate that NASA may not have sufficient capability to carry all the cargo that could be needed on the ISS. The manifests show that, when anticipated growth in national laboratory demands and margin for unforeseen maintenance needs are accounted for, the current number of flights NASA is planning for may not cover all of NASA's anticipated needs beginning in 2014.[25] These shortfalls amount to a total of 2.3 metric tons—approximately the cargo that one SpaceX commercial vehicle can transport to the ISS. NASA has indicated, however, that these shortfalls could increase or decrease because transportation demand, vehicle flight rate, and vehicle capability projections beyond 2015 are estimates only. Additionally, ISS agreements with commercial vehicle providers are not in place beyond 2015 and agreements for use of international partner vehicles beyond 2015 are in negotiations.

Table 3. NASA's Planned Vehicle Launches for 2012 to 2020 to Resupply the ISS

Vehicles	2012	2013	2014	2015	2016	2017	2018	2019	2020	Total
ATV	1	1	1		1			1		5
HTV	1	1	1	1	1	1	1		1	8
SpaceX	3	3	3	2						11
Orbital	2	2	2	2						8
SpaceX Follow On Vehicle A					3	3	2	3	3	14
Orbital Follow On Vehicle B					1	2	3	2	2	10
Total	7	7	7	5	6	6	6	6	6	56

Source: GAO analysis based on NASA documents.

Note: This table does not include flights by the Russian Soyuz or Progress vehicles.

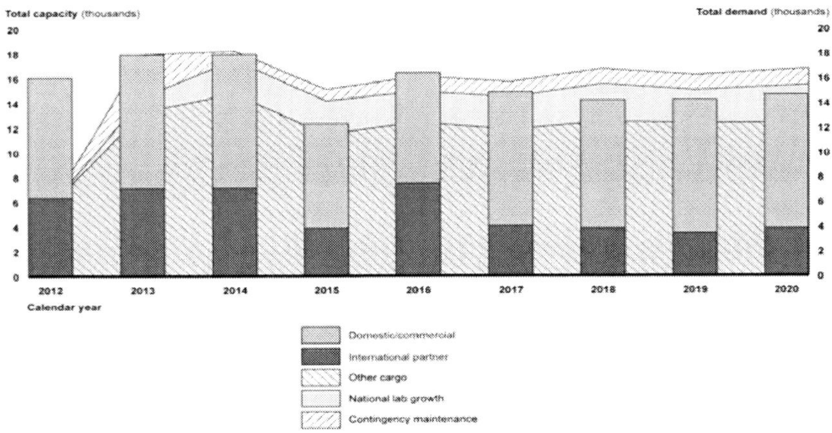

Source: GAO analysis of NASA data.

Figure 4. Cargo Capability of U.S. Commercial and International Partner Vehicles Versus NASA's ISS Sparing Needs from 2012 to 2020.

As we have previously reported, both SpaceX and Orbital are working under aggressive schedules and have experienced delays in completing demonstrations.[26] SpaceX may gain some ground in its schedule by combining demonstration missions planned for 2011, but this likely would not provide much overall schedule gain because previous delays had already placed the program about 18 months behind schedule. Additionally, in the summer of 2011 Orbital experienced an engine failure during testing, which could affect its planned launch vehicle availability. NASA has made efforts to accommodate delayed commercial vehicle development, including use of the final shuttle flight in July 2011 to preposition additional ISS spares. However, if the commercial vehicle launches do not occur as planned in 2012, the ISS could lose some ability to function and sustain research efforts due to a lack of alternative launch vehicles to support the ISS. The need for these vehicles to launch in late 2012 could become even more important as a result of the failure of the Russian Progress re-supply mission in August 2011,[27] one of the last Progress flights that NASA had contracted with Russia to transport cargo to the ISS. As noted by NASA officials, the Russian vehicle was to provide supplies that would help sustain crew and scientific experiments aboard the ISS. At this time, NASA officials indicated that the agency does not believe the loss of the Progress flight increased the need for the commercial launches in 2012 because the agency anticipates that a Progress flight in 2013 will carry the supplies to the ISS and provide margin against delays in the commercial vehicle launches. However, if that Progress flight is not conducted, NASA could become more reliant on the commercial vehicles.

NASA Has Budgeted for Estimated Hardware Spares through 2020

To ensure that spares essential to ISS operations will be available, NASA has budgeted approximately $900 million to supply ISS hardware spares needed to meet functional targets through 2020. NASA officials indicated that, of that amount, the agency plans to spend about $515 million procuring ISS spares to be used from 2010 through 2015 and about $384 million for spares to be used from 2016 through 2020. According to NASA officials, the difference in budgets between the 2010 through 2015 and 2016 through 2020 time frames reflects the need to procure fewer spares from 2016 through 2020 because NASA already purchased and pre-positioned spares aboard the ISS. The officials indicated that this strategy allowed for schedule margin against delayed development of commercial launch vehicles; that is, spares would already be pre-positioned on the ISS in case commercial launch vehicles were not available as planned in 2012. Additionally, the officials stated that the strategy of graceful degradation reduces the amount of spares required as the ISS nears its planned end of life in 2020.

As shown in Table 4 below, of the $515 million planned for spares to be used through 2015, NASA officials anticipate that the agency will spend about $288 million to procure spares for the Environmental Control and Life Support System.[28] The agency officials expect NASA to use the remaining funds, about $227 million, to procure spares for the electrical power system; the thermal control system; communications and tracking systems; command and data handling; and structures and mechanisms, such as hatches and payload racks.

Table 4. Cost Estimates for ISS Hardware Spares to be Used through 2015

System type	Total cost (dollars in millions)	Last delivery date
Electrical Power System	$31.6	2012
Structures and Mechanisms	$92.8	2012
Command and Data Handling	$1.3	2009
Communications and Tracking	$76.5	2013
Thermal Control System	$24.5	2012
Regenerative ECLSS	$233.0	2013
ECLSS	$55.2	2012
Total all systems	$514.9	

Source: GAO analysis based on cost data provided by NASA officials.

Of the $384 million NASA plans to use to procure spares for use in 2016 through 2020, agency officials estimate NASA will need about $196 million to replace ISS batteries and lighting and about $115 million for spares for electrical power, environmental control and life support, and other systems. The remaining budget estimate of about $73 million, as noted by NASA officials, is expected to be used for repair of spares returned to earth via SpaceX commercial launch vehicles. The officials emphasized that budget estimates and hardware quantities for 2016 through 2020 are not firm, and they continually refine both estimates and quantities of spares needed as the agency gains increased knowledge of how long hardware spares will last. For example, officials indicated that they are already adjusting budget projections based on the results of the 2011 functional availability assessment, which indicated that some functions may need more spares than anticipated while other functions may require fewer spares.

NASA IS USING REASONABLE METHODS TO ASSESS STRUCTURES AND HARDWARE

NASA is using reasonable analytical methods to assess structural health as well as station hardware and ORUs that could affect the safety of the station. NASA's software tools, in use since the program's inception, indicate the ISS is structurally sound to support continued operations through 2020 and potentially beyond with some mitigation. Other federal agencies recognize analytical assessments as valid methods for evaluating structural health, and the software NASA uses is widely used by industry. Currently, however, NASA has assessed only 40 percent, by weight, of the assembled ISS because most of the ISS structures have not been on orbit long enough to accumulate the data needed for analysis.[29] NASA expects to complete ISS structural assessments in early 2016. NASA plans to use improved software tools for evaluating life extension of ISS's structures and expects these tools to provide greater insight into the structural health of the ISS. By 2012, NASA expects to complete analytical safety and mission assurance reviews designed to assess operations of the ISS as a whole as well as identify ORUs and other hardware failure of which could result in an increased risk of loss of station or loss of crew. Through 2015, NASA plans to develop methods to mitigate issues identified. When these mitigation actions are developed and implemented, NASA expects to have increased assurance that the ISS can safely operate through 2020 and potentially beyond.

Analysis Ongoing but NASA Expects the ISS to Remain Structurally Sound with Mitigation

Based on prior analysis of structural life usage through 2015 and the robust design of the ISS primary structures, NASA anticipates that—with some mitigation—the ISS will remain structurally sound to support operations through 2020, though its analysis through 2020 is still ongoing. NASA standards for manned spaceflight systems require the ISS program to monitor the structural health of the ISS.[30] Other federal agencies— including the

Federal Highway Administration, Federal Aviation Administration, and National Institute of Standards and Technology— recognize visual, physical, radiographic (x-ray), and sonographic inspections as well as analytical assessments as valid methods for evaluating the structural health of bridges or airplanes. Visual inspections of the ISS are limited to using television cameras to assess the station's exterior, but the cameras do not provide sufficient detail for structural inspection. Additionally, the position of payload racks within the ISS severely hampers visual inspection of the station's interior weld seams. Physical assessments are limited to one ISS structure—the US-funded, Russian-built Functional Cargo Block.[31] These ground-based assessments, conducted on a full-scale model of the cargo block, have shown that it should remain structurally sound through 2028.[32] According to agency officials, NASA has not attempted to develop techniques or equipment to conduct x-ray or sonographic inspection of the ISS. The officials stated that doing so would be extremely expensive, impractical, or both.

Because NASA cannot feasibly use traditional methods to assess the majority of the ISS on orbit, the agency relies on analytical assessments to determine ISS structural health. Since program inception, NASA has used various versions of a software tool known as NASGRO® [33] as its key analytical tool for monitoring the structural health of the U.S. portions of the ISS.[34] NASGRO assumes that non-discernable cracks are present in any given structure and predicts the growth rate and likely size of a crack or "crack growth" based on: (1) the shape of the structure, (2) the material composition of the structure, and (3) the physical loads placed on the structure. NASGRO is widely used by government and industry including the Department of Defense (DOD), Federal Aviation Administration, The Boeing Company, Lockheed Martin, Honda, Airbus, and Spirit Aerosystems. NASGRO is also used by the International Partners, i.e., the Japanese and European space agencies to monitor the structural health of their elements. NASA, in conjunction with other users, continually provides input for the update and release of new versions of NASGRO with increased analysis capabilities and materials knowledge. One NASGRO industry user we talked to said that his company chose to rely on NASGRO as its tool for predicting structural health because it is conservative in its projections, has an extensive materials properties database, and is a Federal Aviation Administration-accepted analysis method for fatigue and damage assessment of structures.

NASA's approach to assessing structural health compares initial NASGRO predictions of crack growth with updated NASGRO predictions of crack growth. NASA used NASGRO to calculate initial predictions of crack growth during the ISS design phase based on the mechanical, thermal and pressure loads NASA expected the ISS to experience throughout its operational life ending in 2015. NASA annually updates crack growth predictions based on the actual loads the ISS has experienced since being placed in orbit.[35] NASA prepares the updated predictions of crack growth with the same version of NASGRO used to prepare the initial prediction. NASA prepared the initial predictions of crack growth for each structural component with the version of NASGRO that was current at that time. As the ISS was developed over several years, NASA uses several versions of NASGRO to assess structural health.

For these comparisons to be useful, the structural elements of the ISS under examination must have been in orbit long enough to accrue meaningful differences between predicted and actual loads. Thus far, NASA has found that the actual loads experienced by ISS structures have been generally less stressful than those originally predicted. The last completed assessment, in fiscal year 2010, examined the ISS structures placed in orbit as of November 2002, which represents approximately 40 percent of the ISS measured by weight.

The structural health assessment uses a risk-based screening approach and examines in detail only those components originally estimated to have less than 6 service lives or safety margins of less than 10 percent.[36] According to program officials, because ISS developers were conservative in their design approaches and in most cases built hardware that exceeded requirements, most structural locations do not need to be examined in detail. As of the last analysis, completed in November 2010, of the 10,000+ structural locations within the on-orbit segments of the ISS, 829 locations were examined in detail because they met screening criteria. Of the 829 locations examined in detail, 196 are using structural life faster than originally predicted, but only 19 no longer meet the original requirement of 60 years of service life. Fifteen of the 19 are located on the P6 truss.

The P6 truss—an end portion of the truss that serves as the structural backbone of the ISS, as shown in figure 5 above—holds 4 of the ISS's 16 solar panels. The P-6 truss provides an example of the impact of actual loads data on the determination of structural health of ISS structures. The truss was left in what was planned as an interim location for an extended period of time following the Space Shuttle Columbia accident. This location left it exposed to greater extremes of hot and cold temperatures than anticipated. According to program officials, the P6 truss may require mitigation to remain sound through 2020 and potentially beyond. NASA is evaluating mitigation processes, such as adding thermal insulation blankets and changing how and where vehicles dock to the ISS, to slow down the rate at which the locations on the P6 truss, as well as the other locations, are expending structural life.

Source: NASA.

Interim location of P6 truss.

Source: NASA.

Permanent location of P6 truss.

Figure 5. P6 Truss in Interim and Permanent Locations.

As part of its life extension effort, NASA is using a single version of NASGRO—NASGRO 6.0—to develop new crack growth baselines for assessing structural health through 2020 and potentially beyond. The agency is deriving the new baselines from the actual on-orbit loads the ISS has experienced since being placed in orbit and predicted loads for using the ISS through 2028.[37] The agency expects to incorporate these new baselines into biennial ISS structural health assessments beginning in 2012. Methodologically, these assessments will be virtually identical to the earlier assessments, but they will use the new baselines and improved NASGRO software with increased capabilities. NASA plans to establish these new baselines in 3 phases, with the last one established in 2015.

- Phase 1, beginning in 2009 and ongoing through 2012, examines structures with original certifications expiring on or before 2013.
- Phase 2, beginning in 2012 and ongoing through 2014, examines structures with original certifications expiring on or before 2017.
- Phase 3, beginning in 2012 and ongoing through 2016, examines structures with original certifications extending beyond 2020.

On the basis of the results of these analyses, NASA will determine what mitigation actions need to be put in place to ensure the ISS remains structurally sound through 2020.

On the basis of familiarity with NASGRO software, NASA officials expect the updated software to provide greater insight into the structural health of the ISS. For example, NASA expects the new version of NASGRO to better capture the impact of high cycle, low impact loads, such as those caused by astronauts running on the treadmill on the structural health of the ISS. Further, NASA expects the new software to support more in-depth analysis of how thermal, pressure and mechanical loads affect ISS structures. NASA also anticipates that improved knowledge of materials properties data within the software will support more informed analysis of how the structures of the ISS respond to different loads and environments.

When NASA begins using the updated software tools, part of the agency's input, the detailed design of the structures, will be based on existing models of the ISS. Officials indicated, however, that in some instances, these models are not detailed enough to take advantage of all the expanded features of the updated NASGRO software. According to the officials, these models were designed to be used with less sophisticated software than NASGRO 6.0. In areas not expected to meet service-life requirements, one of the mitigation strategies available is refined modeling with increased details. For example, more detailed models could allow NASA to more precisely pinpoint structural issues and focus mitigations accordingly.

NASA Is Employing Reasonable Methodologies to Ensure Safety of ISS Hardware through 2020

NASA is employing reasonable analytical methodologies to assess the extent to which ISS hardware, including ORUs and associated components such as flex hoses and tubing, can be safely operated through 2020 and potentially beyond. These methodologies employ established safety and mission assurance practices, governed by agency and program directives,[38] to perform a safety and mission assurance assessment as well as rely on program experts to assess the safety of hardware whose failure could result in the loss of ISS crew or the ISS as a whole.

NASA is performing two types of safety assessments. The first, a safety and mission assurance assessment, is an analytical review of hazards to the ISS as a whole providing a "top-down" look at how extending the life of the ISS through 2020 and potentially beyond affects ISS safety. Additionally, for the second assessment, NASA is using program experts to assess the ORUs[39] and other hardware, whose failure could result in loss of ISS crew or the entire ISS. This assessment is a "bottom-up" analytical review to determine whether this hardware can be safely operated through 2020 and potentially beyond. According to NASA officials, addressing safety from two different perspectives—top down and bottom up—offers increased assurance that safety issues are fully addressed.

NASA's safety and mission assurance practices call for identification of safety hazards and mitigation planning to address hazards. Safety hazards on the ISS are varied and consist of hazards to both the crew and the station. Examples of hazards include loss of crewmembers due to inability to reenter the ISS after conducting a spacewalk or loss of ISS attitude control. As indicated by NASA officials, the agency originally documented these hazards in 92 hazard reports applicable to ISS operations through 2015. NASA is now in the process of conducting a "top-down" review of these reports to determine if existing hazards require additional mitigation as a result of extending ISS life through 2020 and potentially beyond. An official working on the assessment stated that preliminary results of this assessment indicate that 58 hazard reports are affected by extending ISS life beyond 2015 and require further study. Mitigation actions could take a variety of forms, for example, procuring additional ORUs, updating hazard reports and operational procedures, and performing hardware modifications and redesigns.

In conjunction with the "top-down" safety and mission assurance assessment, NASA is using existing hardware teams who are responsible for and knowledgeable of their respective systems to assess the impact of extending station life on ORUs and other hardware. These teams are conducting in-depth, "bottom-up" reviews of ORUs and associated hardware, such as hose assemblies and tubing, to identify hardware that must be replaced before failure because the hardware's failure could lead to loss of crew or station. According to NASA officials, the analysis is currently in progress, but preliminary results indicate that some hardware will not be affected by ISS life extension, while other hardware may require replacement or other mitigation to ensure it does not fail and result in a catastrophic loss of life or station. For example, NASA's early findings show that 52 hardware items, including tubes and hose assemblies in the Internal Thermal Control System, the system that moderates temperature throughout the ISS, can be used until they fail because their failure would result only in degradation of system functions and not endanger the ISS or crew.

NASA expects to complete both the "top-down" safety and mission assurance and "bottom-up" hardware assessments by 2012. Agency officials stated that they will begin mitigation planning immediately after the completion of both assessments and continue that planning through 2015. They noted that the ISS program will implement corrective actions as mitigation plans are developed and put in place, so that some mitigation efforts will occur as early as 2012 and some after 2015. When mitigation actions are developed and in place, NASA expects to have increased assurance that the ISS can safely operate through 2020 and potentially beyond.

CONCLUSION

Extending the operational life of the ISS to 2020 will allow NASA and the United States to capitalize on the nearly two decades and billions of dollars invested in its construction and operation. During the time required to finish ISS construction, however, many changes occurred in the assumptions and conditions underlying NASA's initial ISS supportability strategy—the planned operational life of the ISS has extended, the space shuttle program has ended, and ORUs on the ISS are lasting longer than originally expected. In response to the changing conditions, NASA has appropriately evolved its strategy for supporting the ISS through 2020. Assessing ISS extended life necessarily requires the use of sophisticated analytical techniques and judgments. To that end, the agency has adopted reasonable approaches and methodologies for estimating the spares needed for continued operation and ensuring that the ISS structure and hardware are safe for operation through 2020 and potentially beyond. For example, NASA actively incorporates new knowledge to update estimates of MTBF for its ORUs, which provides greater efficiency in the use of resources necessary for purchasing spare parts. While this approach is sound, it is important that this process continue to evolve as the agency gains knowledge to ensure the most effective and efficient use of resources for ISS utilization through 2020. NASA has not indicated, however, that it plans to incorporate this new knowledge by revisiting the variance parameter that governs the relative weight given to on-orbit experience versus manufacturers' original reliability estimates in conducting Bayesian updates. Doing so when the appropriate amount of knowledge is obtained could allow NASA to achieve even greater efficiency in the purchase of spares. In considering these assumptions, however, NASA must balance efficiency with risk. If NASA weights on-orbit experience too heavily, it risks overstating the MTBF and potentially underestimating the number of spares needed, especially for ORUs that have not experienced a random failure. Conversely, weighing manufacturers' original estimates too heavily could result in NASA expending resources on unneeded spares. Moving forward, NASA needs to carefully balance these competing information sources based on future on-orbit experience to ensure an appropriate risk posture in the procurement of spares. The basis underlying NASA's functionality targets and goals, which are revisited annually, could also be affected by factors such as increased knowledge of performance and functionality and additional capabilities provided as new launch vehicles become available. Understanding the rationale behind NASA's functionality targets and goals is important so that as new information is gained it can be assessed to ensure that it remains valid. For example, if a particular goal was set at 98 percent due to the assumed availability of spares and those spares become unavailable, that goal may no longer be achievable and mitigations or an alternative approach would need to be devised to meet that goal. Without documentation of the rationale behind the original decision, ISS officials have limited means to assess whether the rationale remains valid and the goal is achievable.

Recommendations for Executive Action

As the ISS program accumulates additional knowledge about the on-orbit performance of ORUs, we recommend that the NASA Administrator direct the ISS program manager to revisit, as appropriate, the relative weight given to manufacturers' original reliability estimates and to the actual reliability of ORUs based on on-orbit experience by reexamining the program's choice of a value for the parameter that governs the variance of the original MTBF's probability distribution.

To ensure decision makers have a full understanding of the rationale behind ISS functionality targets and confidence levels to better inform future decisions, we recommend that the NASA administrator direct the ISS program manager to ensure these rationales are documented, at a minimum, in the minutes from meetings of the Space Station Program Control Board.

Agency Comments and Our Evaluation

In written and oral comments on a draft of this report, NASA concurred with our recommendations. NASA acknowledged the benefit of revisiting the relative weight given to manufacturers' original reliability estimates and to the actual reliability of orbital replacement units and indicated that the ISS program office intends to reexamine these weighting values as appropriate. NASA also acknowledged that the ISS program office should document the rationale underlying the Program Manager's decisions concerning ISS functionality targets and confidence levels and indicated that the ISS Program Manager is already in the process of implementing this recommendation. Separately, NASA provided technical comments, which have been addressed in the report, as appropriate.

Cristina T. Chaplain
Director
Acquisition and Sourcing Management

APPENDIX I: SCOPE AND METHODOLOGY

We interviewed and obtained briefings and relevant documents from knowledgeable National Aeronautics and Space Administration (NASA) and contractor officials on the content and decision-making approach NASA used to prepare its assessment in response to section 503 of the NASA Authorization Act of 2010.[1] directing the agency to assess its plans for ensuring essential spares were available to support the International Space Station (ISS) through the station's planned operational life. We also reviewed the scope and methodology NASA used in the ISS assessment. We conducted our work at NASA headquarters in Washington, D.C., and NASA's Johnson Space Flight Center, in Houston, Texas.

For purposes of assessing NASA's findings and approach to estimating spares essential for ISS operation through 2020, we focused our efforts on the statistical techniques NASA used to calculate operational mean time between failure (MTBF) rates and the modeling techniques NASA uses to assess functional availability as well as NASA's process for establishing functionality targets and confidence levels for ISS functional systems. For a limited number of orbital replacement units (ORU), we recalculated the values NASA obtained for the operational MTBF based on its statistical methodology. We also conducted limited tests of the data NASA used to support the functional availability model. We discussed the ISS program's approach with NASA experts outside the ISS program at NASA headquarters. We discussed NASA's approach for ensuring data reliability and configuration control of ISS models and analytical systems with cognizant and responsible officials. We also obtained and reviewed a briefing on the results of an internal review of ISS program models and analytical systems. We limited the scope of our assessment to the scope of NASA's report, i.e., the sparing necessary to support critical functions as modeled in the ISS functional availability assessment. We did not examine the sparing or launch vehicle needs of the international partners or the sparing needs of functions not included in the ISS functional availability assessment.

For purposes of assessing NASA 's findings and approach to monitoring and assessing the structural health of the ISS, we interviewed and obtained briefings and relevant documents from knowledgeable NASA and contractor officials at the Johnson Space Center. We obtained and reviewed the results of NASA's past assessments of ISS primary structures to identify issues that could affect ISS operations through 2020. We obtained and reviewed Department of Defense and NASA guidance documents and determined

NASA's basis for establishing four service lives as its baseline requirement for structural life. We discussed NASA's basis for establishing six service lives as the cutoff point for conducting ongoing structural health assessments with knowledgeable agency and contractor officials. Through discussions with NASA's NASGRO experts at Johnson Space Center; with NASGRO experts responsible for managing the commercial side of the NASGRO structural analysis software at Southwest Research Institute; and commercial users of NASGRO, we obtained an understanding of NASA's structural heath assessment software tools and models, how NASA uses them to support its assessments of structural health, and NASA's plan for incorporating improved software tools into future assessments. We reviewed Federal Aviation Administration and Federal Highway Administration policies guiding the inspection of aircraft and bridges and compared them to NASA's approach to measuring structural health. We identified NASA's plans to use improved software with increased knowledge of material properties and enhanced software tools to gain improved measurements of structural health. We discussed with knowledgeable NASA officials the mitigation strategies NASA is considering to address structural problems discovered by structural health assessments.

To determine the viability of NASA's findings and plans regarding transportability for and supportability of the ISS, we examined NASA's plans for transporting cargo to support the ISS through 2020 and examined ISS launch vehicle schedules from 2012 through 2020. We discussed those schedules, including launch vehicle availability, with NASA officials. We also reviewed NASA launch manifests for 2012 through 2020. We compared the launch schedules to the planned cargo manifests and determined whether planned launch vehicles would be capable of meeting ISS supportability needs. Additionally, we reviewed prior GAO reports and testimonies addressing commercial domestic launch vehicle development and NASA's management and program challenges. To evaluate the status of NASA's efforts to estimate the costs of ISS spares through 2020, we obtained documents providing information about NASA's past and predicted ISS sparing costs. We discussed these estimates with knowledgeable NASA officials to gain insight into how the cost estimates were developed and how changes in cargo needs and vehicle availability might affect the estimates.

To assess NASA's findings and plans for ensuring ISS safe operations through 2020, we interviewed and obtained briefings from NASA officials on their methodologies for assessing the safety of the ISS as a whole as well as the hardware that could affect safety of the crew or station through 2020.

Specifically, we reviewed the methodology NASA is using to perform top-down safety and mission assurance assessments. We also reviewed the methodology NASA is using to perform bottom-up assessments of ISS hardware to determine if there are ISS hardware safety issues associated with extending ISS mission life. Additionally, we assessed NASA's approach to reviewing ISS safety in terms of the agency's and Johnson Space Center's internal regulations and guidance. We also discussed NASA's approach with ISS program and contractor officials. We interviewed and obtained briefings from NASA officials on the results-to-date of ISS safety and mission assurance and hardware assessments and discussed issues that could affect the safety of ISS operations through 2020 as well as NASA's plans for mitigating any such issues.

We conducted this performance audit from February 2011 to November 2011 in accordance with generally accepted government auditing standards. Those standards require that we plan and perform the audit to obtain sufficient, appropriate evidence to provide a reasonable basis for our findings and conclusions based on our audit objectives. We believe that the evidence obtained provides a reasonable basis for our findings based on our audit objectives.

End Notes

[1] Pub. L. No. 111-267.

[2] GAO, *International Space Station (ISS): Ongoing Assessments for Life Extension Appear to be Supported*, GAO-11-519R (Washington, D.C.: Apr. 11, 2011).

[3] Four service lives is a NASA and Department of Defense standard for spacecraft. According to agency officials, most components are designed to operate beyond four service lives.

[4] The structures of the ISS encounter thermal, pressure and mechanical loads. According to program officials, thermal loads are developed due to temperature gradients across or through a structure and/or by a mismatch in materials at a joint; pressure loads are developed due to pressure differences on the inside versus the outside of a structure, such as the atmosphere in a crew compartment versus the vacuum of space; and mechanical loads are developed when a force is applied to the structure, such as vehicle dockings and undockings.

[5] National Aeronautics and Space Administration Authorization Act of 2008, Pub. L. No. 110-422, § 601; National Aeronautics and Space Administration Authorization Act of 2010, Pub. L. No. 111-267, § 503.

[6] In 2004, President George W. Bush established a new space exploration policy-*A Renewed Spirit of Discovery*: The President's Vision for U.S. Space Exploration (Vision)- which called for the retirement of the space shuttle and development of a new family of exploration systems to facilitate a return of humans to the moon and eventual human spaceflight to Mars.

[7] Department of Defense, *Military Handbook: Reliability Prediction of Electronic Equipment*, MIL-HDBK-217F (Washington, D.C.: Feb. 28, 1995).

[8] Bayesian analysis is a method of statistical inference (named for English mathematician Thomas Bayes) that allows one to combine prior information about a population parameter with evidence from information contained in a sample to guide the statistical inference process.

[9] The ISS Reliability and Maintainability Panel at Johnson Space Center in Houston, Texas, provides support to all ISS hardware teams at all NASA centers.

[10] S-band is a radio band used by NASA to communicate with the ISS.

[11] Monte Carlo modeling is used to approximate the probability outcomes of multiple trials by generating random numbers. Monte Carlo models are widely used throughout government and industry to calculate approximate probabilities for complex events when the calculation of exact probabilities would be difficult or impossible.

[12] A general description of a probability distribution is a function that describes the probability of a random variable taking certain values.

[13] A probabilistic risk assessment is a comprehensive, structured, and logical analysis method aimed at identifying and assessing risks in complex technological systems for the purpose of cost-effectively improving their safety and performance in the face of uncertainties. Probabilistic risk assessments assess risk metrics and associated uncertainties relating to likelihood and severity of events adverse to safety or mission. (NASA Procedural Requirements 8715.3C, *NASA General Safety Program Requirements*).

[14] NASA Procedural Requirements 8705.5A, *Technical Probabilistic Risk Assessment (PRA) Procedures for Safety and Mission Success for NASA Programs and Projects.*

[15] Because the failure rate is constant, the Bayesian estimation process does not care whether the on-orbit run time is a single ORU operating for 100,000 hours, or 4 identical ORUs operating for 25,000 hours.

[16] The ISS program maintains an "original MTBF" data field in its Modeling and Analysis Data Set (MADS)—the database it maintains on key orbital replacement unit parameters—and uses the information from this data field in its Bayesian estimation calculations. We discussed with ISS program representatives the steps they had taken to ensure that the MTBF data captured in MADS reflect the data from the reliability analysis performed by the original equipment manufacturers. We also confirmed for a small number of orbital replacement units, that the MTBF data contained in MADS matched the MTBF data from the original equipment manufacturers' reliability analysis. However, because many of these reliability reports were performed well over a decade ago, and many exist only in hard copy at a long-term storage facility, we did not conduct detailed tests of these underlying data.

[17] NASA, *Space Station Freedom: External Maintenance Task Team Final Report.* (July 1990). (This study is more commonly referred to as the Fisher-Price report, after its two principal authors.)

[18] According to ISS program office representatives, the July 2010 failure of the pump module assembly has been categorized as a random failure. However, if further analysis and testing reveal that the unit failed for some other reason—for example, a workmanship problem—then the failure would be re-categorized, and the Bayesian estimation calculation would be repeated without the failure.

[19] The functional availability risk assessment employs a Monte Carlo simulation model to calculate the number of spares required to meet the functionality targets and confidence goals for each function. Monte Carlo modeling is widely accepted by industry and government as appropriate for these types of simulations.

[20] The 2010 functional availability assessment had focused on the number of spares required to meet the functionality target and confidence goals through 2020. The 2011 functional availability risk assessment addressed meeting functionality targets and confidence goals beyond 2020. However, we did not receive information on the 2011 assessment until late in our review and, therefore, did not have time to assess the results.

[21] GAO, *Standards for Internal Control in the Federal Government*, GAO/AIMD-00-21.3.1 (Washington, D.C.: November 1999).

[22] In 2008 and 2009, the ATV and HTV vehicles respectively flew to the ISS and docked at the station to demonstrate their capabilities. In 2011, both vehicles again launched. These flights were the second for both systems.

[23] GAO, *NASA: Commercial Partners Are Making Progress, but Face Aggressive Schedules to Demonstrate Critical Space Station Cargo Transport Capabilities*, GAO-09-618 (Washington, D.C.: June 16, 2009).

[24] NASA has awarded contacts to SpaceX and Orbital for cargo resupply services to the ISS through 2015. Planned follow-on vehicles are the vehicles NASA will use for flights beyond those currently on contract.

[25] NASA selected a nonprofit organization to facilitate increased use of the ISS on-orbit research capabilities by entities outside of NASA.

[26] GAO, *NASA: Commercial Partners Are Making Progress, but Face Aggressive Schedules to Demonstrate Critical Space Station Cargo Transport Capabilities*, GAO-09-618 (Washington, D.C.: June 16, 2009).

[27] The Progress vehicle's mission failed when the third stage engine failed during launch.

[28] The Environmental Control and Life Support System (ECLSS) makes the ISS habitable for crew by providing or controlling oxygen supply, fire detection and suppression, waste management and water supply. The total of $288 million includes $233 million NASA has budgeted to procure spares for the Regenerative ECLSS, which recycles liquid waste into potable water and oxygen. In terms of estimating the needed spares, the Regenerative ECLSS is the most high-risk functional system, as several of its ORUs are failing faster than originally anticipated.

[29] The timing of initial structural assessments of ISS components was originally driven by programmatic judgment calls as to when predicted and actual loads had diverged enough to provide a meaningful assessment. The timing and order of current assessments are now driven by the assembly sequence of the ISS and NASA's desire to complete the initial assessment of each component before its 15-year structural certification expires.

[30] NASA Technical Standard-5007, *General Fracture Control Requirements for Manned Spaceflight Systems* (March 13, 2001); and NASA Technical Standard-5003, *Fracture Control Requirements for Payloads Using the Space Shuttle* (October 7, 1996) establish the requirements for protection against fracture within spaceflight system structures. Fractures, simply put, are cracks that form in structures as a result of the loads they encounter. Fracture criticality deals with the likelihood that a structural element will fail under the loads it experiences. For a component to be considered fracture critical, its potential failure must cause loss of life, loss of mission, or loss of a manned spaceflight system. Fracture control ensures a paper trail of fracture analysis and non-destructive testing/evaluation.

[31] The Functional Cargo Block (FGB) was the first portion of the ISS put into orbit.

[32] The Russian Space Agency used repetitive physical tests of a full-scale model of the FGB to initially certify the FGB for 15 years of life on-orbit. The original certification performed by the Russian Space Agency was set to expire in 2013. To assess the viability of using the FGB an additional 15 years to 2028, NASA relied on the Russian Space Agency to conduct

further testing on the full-scale model originally used to certify the FGB. In July 2011, NASA and the Russian Space Agency recertified the FGB for 15 additional years of service—until 2028.

[33] NASGRO is a suite of programs used to: analyze fatigue crack growth and fracture, perform assessments of structural life, process and store fatigue crack growth properties, and analyze fatigue crack formation.

[34] When conducting structural assessments of the ISS, NASA assesses each structural component with the same version of NASGRO.

[35] Actual loads are calculated by analytical modeling of the ISS structure. The models are validated by data obtained from sensors on-board the ISS.

[36] NASA engineers believe there is a reasonable expectation that components that have analytically demonstrated more than 6 service lives, or 90 years, are not likely to fail within the original service life timeframe.

[37] NASA is conducting life extension structural and hardware analyses through 2028 for two reasons: (1) the physical tests of the FGB assessed its structural soundness through 2028 and, (2) the program does not want to have to replicate the analysis if the ISS's life is extended beyond 2020.

[38] These directives include NM 7120-97, NASA Procedural Requirement 7120.5D, SRB Handbook, *NASA Space Flight Program and Project Management Requirements*, which is the interim directive to NPR 7120.5D; NASA Policy Directive 8700.1E, *NASA Policy for Safety and Mission Success*, (October 28, 2008); NASA Policy Directive 8700.3B, *Safety and Mission Assurance Policy for NASA Spacecraft, Instruments, and Launch Services*, (October 28, 2008); Space Station Program 30599 Revision A, *Safety Review Process* (January 11, 1995); Space Station Program 30309 Revision E, *Safety Analysis and Risk Assessment Requirements Document*, (October 28, 1994).

[39] NASA's safety assessment covers only those ORUs whose failure could result in loss of life or loss of station. These ORUs are a subset of those covered by NASA's approach to determining spares required to ensure utilization of the ISS through 2020 discussed earlier in the report.

End Note for Appendix I

[1] Pub. L. No. 111-267.

INDEX

A

B

C

D

E

N

O

P

T

U

V

W